Complexity and Complex Ecological Systems

Complexity and Complex Ecological Systems

Stanislaw Sieniutycz
Professor of Chemical Engineering, Warsaw University of Technology, Faculty of Chemical and Process Engineering, Poland

ELSEVIER

Elsevier
Radarweg 29, PO Box 211, 1000 AE Amsterdam, Netherlands
The Boulevard, Langford Lane, Kidlington, Oxford OX5 1GB, United Kingdom
50 Hampshire Street, 5th Floor, Cambridge, MA 02139, United States

ISBN: 978-0-443-19237-1

Publisher: Candice Janco
Acquisitions Editor: Anita Koch
Editorial Project Manager: Emerald Li
Production Project Manager: Bharatwaj Varatharajan
Cover Designer: Miles Hitchen

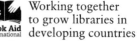

Typeset by STRAIVE, India

Contents

Preface

The book on *Complexity and Complex Ecological Systems* is the third volume on complexity and systems written by the same author. The first book, *Complexity and Complex Thermo-Economic Systems*, published at the beginning of 2020, focused on thermodynamic and economic aspects of complexity and complex systems. The second book, *Complexity and Complex Chemo-Electric Systems*, published early in 2021, evaluated chemical, bio-electrochemical, and electro-mechanical aspects of complex systems. This third book is devoted to ecological systems. The book transfers the reader into the rich world of ecological and environmental problems involving such basic topics as thermodynamic aspects of ecology, diversity and spread of populations, trophic relationships of coastal and estuarine ecosystems, dynamical food webs, inference from computer simulations, mathematical ecology, and optimization of ecological processes.

Recent years have brought new and unexpected problems. In spite of moderate medical progress in overcoming the coronavirus epidemic, we still have to live and work in a world partly fragmented by this threat while suffering from a multitude of social problems different than those ordinary humans could expect several years ago. The structure and change in human decisions should involve not only variables related to life today such as health and medical recommendations, but also take into account new problems, such as those resulting from the rapid change of the social or political situation. We need more and more help from organizations, societies, the environment, science, medicine, etc., yet the range of this help may differ today from what was required previously. This implies the essential role of the knowledge related to environmental, social, and medical information that may prevail for at least a period of time over the role of more traditional information stemming from exact sciences and technologies.

It is familiar to the reader that energy and matter balances, the results of invariance of some basic (action) principles in physics, constitute macroscopic models describing our macroscopic world. These balances incorporate limitations on the working parameters of real processes, which should be taken into account when formulating constraints on the performance of macroscopic systems. Thermodynamics, kinetic theory, and/or some basic experiments provide data on the static and transport properties needed in calculations. These data are necessary to express conservation laws in terms of variables used in system modeling, analysis, and synthesis. After making the selection of state variables, controls, and parameters, we can arrive at an optimization model reflecting an optimization problem. In fact, macroscopic variables are frequently state and control coordinates in various optimizations. Moreover, suitable macroscopic variables may also constitute the performance criteria of accepted optimizations. This stage may also influence the formulation of paraeconomic optimization criteria governing thermo-economic optimization of systems, as, for example, those in our first complexity book.

Briefly, the basic purpose of this book is to investigate formulations, solutions, and applications describing the optimal performance of selected ecological systems, including those considered from the standpoint of a prescribed optimization criterion (complexity measure, chemical production, power yield, topological structure, economical profit, discretization scheme, iterative strategy, simulation time, convergence test, technical index, etc.).

Chapter 1 characterizes the development of ecology and the onset of ecological optimization before the year 2000. This involves considerations involving ecological performance functions set before 2000. They are usually linked with thermodynamic optimization. The thermodynamic ideas contained in the popular books of Prigogine (Glansdorff and Prigogine, 1971) and in the later books along this direction (Nicolis and Prigogine, 1977) diffuse quickly to the ecology and soon receive their own characteristic pictures. Prigogine is best known for extending the second law of thermodynamics to systems that are far from equilibrium, and implying that new forms of ordered structures could exist under such conditions. He called these forms "dissipative structures," pointing out that they cannot exist independently of their environment. Nicolis and Prigogine (1977) showed that the formation of dissipative structures allows order to be created from disorder in nonequilibrium systems. These structures have since been used to describe many phenomena of biology and ecology. The ecological core was pointed out in some books and papers, such as Ulanowicz (1997). It was then shown that the simplest ecosystems can display dynamical behavior unpredictable in certain situations. In certain cases, when riddled basins are found, even qualitative predictability was denied.

Chapter 2 continues the history of the further development of thermodynamic approaches to ecology. Thermodynamics is used increasingly in ecology to understand the integral properties of ecosystems because it is the basic science that describes energy transformation from a holistic viewpoint. Many thermodynamically oriented contributions to the ecosystem theory appear; therefore, an important current step toward integrating these contributions is to present them synthetically (Jørgensen, 2001). An ecosystem consists of interdependent living organisms that are also interdependent with their environment, all of which are involved in a continual transfer of energy and mass within a general state of disequilibrium. Further on, considerations are devoted to complex states and complex transformations in ecosystems as well as in growth and aging phenomena. Schrödinger's important 1967 paper shows that living systems seem to defy the second law of thermodynamics because, within closed systems, entropy should be maximized and disorder should reign. Living systems, however, are an antithesis of such disorder. They display excellent levels of order created from disorder. For instance, plants are highly ordered structures that are synthesized from disordered atoms and molecules found in atmospheric gases and soils (Schneider and Kay, 1995). By turning to nonequilibrium thermodynamics, Schrödinger (1967) first recognized that living systems exist well in a world of energy, material, and flows (fluxes), whereas the associated complexity of biology leads to the creation of order from disorder (Schrödinger, 1967; Schneider and Kay, 1995).

In Chapter 3, we describe some meaningful properties of evolving complex systems. A valuable contribution displaying diverse properties of ecological systems is due to Ulanowicz (1997, 2000). His ecological performance functions, initially linked with thermodynamic optimization criteria, expanded in time into more complex and sophisticated structures describing ecosystems more adequately and with more versatility. A phenomenon essential to ecology is that of ecosystem succession, or a more or less repeatable temporal series of configuration that an ecosystem will assume after a major disturbance or upon the appearance of a new area of the given habitat. Initially, succession was described in terms of natural history but more recently ecologists have attempted to describe succession, or system development, more formally (Odum, 1969). The goal in quantitative ecology is to describe the succession process in purely numerical terms.

Chapter 4 treats genetic diversity and the spread of populations. This chapter follows in large part the recent results of researchers affiliated with Ghent University (Mortier et al., 2020), who have shown that by extending the original diversity ideas of Pielou (1969), environmental change can move the physiological limits of a range. Therefore, this can lead to range expansions as determined by population growth and that spread ranges can alternatively expand beyond the existing geographical limits by the introduction of individuals away from their original range. But alongside the environmental opportunities for range expansions, population spread requires individual capabilities. Individual-level life history traits related to reproduction and dispersal will influence the extent and variation in population spread and of range border dynamics. As these traits have a genetic basis in many organisms, range dynamics should to an important extent be determined by the population's genetic composition. Genetic diversity, in numbers and in variation in the identity of genotypes, has a well-studied positive effect on various ecological processes. Genetic diversity tends to improve ecological performance, as expressed by fitness-associated proxies as higher population growth rates. This positive relationship between genetic diversity and a variety of demographic processes can be explained by several mechanisms.

Chapter 5 reviews statistical inference for complex computer models, especially ecological ones. This chapter follows the recent results of Oberpriller et al. (2021), who first formulated and investigated a consistent statistical theory for robust inference in complex computer ecological simulations. Ecological systems are often complex and interdependent. To understand these systems, and to forecast their dynamics under changing conditions, ecologists rely increasingly on complex computer simulations. The synonymous terms they use include process-based models, mechanistic models, and system models. For any of these models, precise forecasts and correct estimates of predictive uncertainty are paramount for their scientific interpretation. The goal of this chapter is to explore these problems in detail and provide on this basis an overview of strategies for robust statistical inference with complex computer simulations of ecosystems.

Chapter 6 characterizes biodiversity maintenance in coastal and estuarine ecosystems, and describes approaches that scientists and managers have taken to classify estuaries and near-shore coasts. The classification of estuaries and coasts is motivated by different perspectives (e.g., landform geomorphology, evolutionary origins, and formative processes), purposes (e.g., understanding structure, variability and dynamics, functions and values, and interaction with adjoining fluvial and coastal ecosystems), and applications (e.g., categorizing, mapping, and management) over diverse temporal and spatial scales. As experienced throughout the history of most sciences, the classification of diverse objects is a foundational step in the progress and application of a discipline, and particularly so in applied sciences. Appreciating how different classes of estuaries and coasts evolve and function is a prerequisite to identifying the approaches and tools needed for management issues and drivers and to identifying and predicting change and assessing impacts. However, as useful as these disciplinary classifications are, interdisciplinary classifications remain elusive, especially those linking the geomorphic form and physical structure and dynamics with ecological and water quality or broader ecosystem functions, goods, and services. Ecological models show, among others, that complexity usually destabilizes food webs.

Chapter 7 describes the study of food webs, a central theme within ecology for decades. Their data of structure and dynamics are used to assess a range of key properties of communities. Yet, many food webs are sensitive to sampling effort, which is rarely considered. Further, most studies have used either species- or size-averaged data for both nodes and links, rather than individual-based data. This is the level of organization at which trophic interactions occur. This practice of aggregating data hides a considerable amount of biologically meaningful variation and could, together with potential sampling effects, create methodological artifacts. New individual-based approaches could improve our understanding of, and ability to predict, food web structure and dynamics. To understand the mechanism driving the complexity and stability of food webs, an eco-evolutionary food web model is developed in which new species emerge via evolution from existing ones and ecological interactions determine which species are viable and which go extinct. The food-web structure emerges from an interplay between speciation and trophic interactions.

Chapter 8 analyzes Pielou's book on mathematical ecology, reach in new ideas and knowledge. The purpose of this book is to serve as a text for mature students and to demonstrate the wide array of ecological problems that invite continual investigation. Various professionals are chemical engineers, physiologists, medical engineers, control engineers, environmental engineers, pharmacologists, etc. The literature of optimizations in the chemical and economical worlds keeps growing, so that efficient literature searches are still necessary. The book is intended as a collection of chapters addressed to actively working scientists and students. The book is basically a textbook in the feature of a reference book. Because of the abundance of literature discussion and the numerous references, the book should constitute a valuable and helpful reference volume for any reader. The book can be used in schools, libraries, and Internet centers as a directory for a guided tour described in the book chapters, each tour representing a research problem.

Chapter 9 is titled *Optimizing in Ecological Systems.* It offers a self-contained analysis of Pontryagin's maximum principle and related criteria of dynamic optimization. The problem formulation is detailed and involves both continuous and discrete models. Detailed are also mathematical conditions of optimality. Because of the presence of a sufficient amount of vast theoretical information, it is not assumed that using other sources will be necessary when optimizing typical problems of ecology. In the most difficult optimization problems, complete analyses should be developed to attain suitable solutions to optimization models arising in ecology.

Considering the goals of the book, one can ask: How could readers (academically/ professionally) profit from reading this book? The answer is as follows:

When this ecological volume is used as a textbook, it can provide basic or supplementary material in the following courses, conducted mostly in engineering and ecological departments of universities:

- Engineering and ecological systems (undergraduate)
- Mathematical description of biological systems (undergraduate)
- Modeling and optimal control in ecological units (graduate)
- Anode-supported solid oxide fuel cells (SOFCs) for conservation of ecological life (graduate)
- Diversity of ecological systems: theory and practice (graduate)
- Computational aspects of ecological modeling (graduate)

Having read the book, the reader will gain the necessary information on what has been achieved to date in the field of complexity and complex ecosystems, what new research problems in ecology could be stated, or what kind of further studies should be developed within the specialized modeling of complex ecological systems. It is expected that the information contained in the book will help to improve both the abstract and technical skills of the reader. The book is especially intended to attract graduate students and researchers of environmental departments (yet possibly including chemical engineering departments). The author hopes that the book will also be a helpful source to actively working professionals, engineers and students.

Finally, we would like to list other related books that can target a similar audience with this type of content. Elsevier has to date published the following set of application-oriented books:

Variational and Extremum Principles in Macroscopic Systems (Ed. by S. Sieniutycz and H. Farkas, Elsevier, Oxford, 2000).
Energy Optimization in Process Systems (by S. Sieniutycz and J. Jeżowski, Elsevier, Oxford, 2009).
Energy Optimization in Process Systems and Fuel Cells (by S. Sieniutycz and J. Jeżowski, Elsevier, Oxford, 2013 (2nd ed.).
Thermodynamic Approaches in Engineering Systems (by S. Sieniutycz, Elsevier, Oxford, 2016).
Optimizing Thermal, Chemical and Environmental Systems (by S. Sieniutycz and Z. Szwast, Elsevier, Oxford, 2018).
Energy Optimization in Process Systems and Fuel Cells (by S. Sieniutycz and J. Jeżowski, Elsevier, Oxford, 2018 (3rd ed.).
Complexity and Complex Thermo-Economic Systems (by S. Sieniutycz, Elsevier, Oxford, 2020).
Complexity and Complex Chemo-Electric Systems (by S. Sieniutycz Elsevier, Oxford, 2021).

Reading *Complexity and Complex Ecological Systems* will provide the opportunity for treating all the above books because of their unity of teaching style as a teaching cluster.

References

Glansdorff, P., Prigogine, I., 1971. Thermodynamic Theory of Structure Stability and Fluctuations. Wiley, New York.

Jørgensen, S.E. (Ed.), 2001. Thermodynamics and Ecological Modelling. Lewis Publishers of CRC Press, Boca Raton.

Mortier, F., Masier, S., Bonte, D., 2020. Genetically diverse populations spread faster in benign but not in challenging environments. Ecology 102 (6), e03345. https://doi.org/10.1002/ecy.3345.

Nicolis, G., Prigogine, I., 1977. Self-Organization in Nonequilibrium Systems. Wiley, New York.

Oberpriller, J., Cameron, D.R., Dietze, M.C., Hartig, F., 2021. Towards robust statistical inference for complex computer models. Ecol. Lett. 24 (6), 1251–1261.

Odum, E.P., 1969. The strategy of ecosystem development. Science 164, 262–270.

Pielou, E.C., 1969. An Introduction to Mathematical Ecology. Wiley Interscience, New York.

Schneider, E.D., Kay, J.J., 1995. Order from disorder: the thermodynamics of complexity in biology. In: Murphy, M. P., O'Neill, L.A.J. (Eds.), What Is Life: The Next Fifty Years. Reflections on the Future of Biology. Cambridge University Press, Cambridge, UK, pp. 161–172.

Schrödinger, E., 1967. What Is Life? Cambridge University Press, Cambridge, UK (first ed. in 1944).

Ulanowicz, R.E., 1997. Ecology, the Ascendent Perspective. Columbia University Press, New York, p. 201p.

Ulanowicz, R.E., 2000. Ascendancy: a measure of ecosystem performance. In: Jorgensen, S.E., Muller, F. (Eds.), Handbook of Ecosystem Theories and Management. Lewis Publishers, Boca Raton, USA, pp. 303–315.

Acknowledgments

The author started his research and collecting scientific materials on ecosystem theory and complexity principles during his stay at the Chemistry Department of The University of Chicago and then while lecturing a course on the ecosystem theory for students of Faculty of Chemical Engineering at the Warsaw University of Technology (Warsaw TU). An early part of suitable materials was obtained in the framework of two national grants, namely, Grant 3 T09C 02426 from the Polish Committee of National Research (KBN). In preparing this volume, the author received help and guidance from Marek Berezowski (Faculty of Engineering and Chemical Technology, Cracow University of Technology), Andrzej Ziębik (Silesian University of Technology, Gliwice), Andrzej B. Jarzębski (Institute of Chemical Engineering of Polish Academy of Science and Faculty of Chemistry at the Silesian TU, Gliwice), Elżbieta Sieniutycz (University of Warsaw), Lingen Chen (Naval University of Engineering, Wuhan, P.R. China), and Piotr Kuran (Faculty of Chemical and Process Engineering at the Warsaw University of Technology). The author is also sincerely grateful to Piotr Juszczyk (Warszawa TU) for his careful and creative work in making all necessary artwork for this book. An important part of preparing any book is the process of reviewing; thus, the author is very much obliged to all researchers who patiently helped him to read through subsequent chapters and who made valuable suggestions. The author, furthermore, owes a debt of gratitude to his students who participated and listened to his lectures on complexity and complex ecosystems in the period 2010–2016. Finally, appreciation also goes to Anita Koch, Elsevier's Acquisition Editor, and the whole book's production team in Elsevier for their cooperation, help, patience, and courtesy.

CHAPTER 1

Early works in ecology and ecological approaches

1.1 Introduction

The purpose of this chapter is to outline the development of ecology and ecological optimization before the year 2000. Considerations involving ecological performance functions set before the year 2000 are usually linked with thermodynamic optimization see: Angulo-Brown, 1991; Ayres, 1978, 1999; Ayres et al., 1996, 1998; Faber et al., 1996; Kay and Schneider, 1992b; Odum, 1971, 1988; Szargut, 1986, 1990; Valero, 1995, 1996; Yan, 1993, and others.

The applications stretch from the realm of thermal technology to chemical and other technologies(Szargut et al., 1988). Particularly fruitful are ecological applications of exergy in their analyses of cumulative exergy consumption and cumulative exergy losses (Szargut, 1990) (Figs. 1.1 and 1.2).

Thermodynamic ideas expressed by Prigogine (1961) and his co-workers, Glansdorff and Prigogine (1971) and Nicolis and Prigogine (1977), diffused quickly to ecology which soon received its characteristic thermodynamic picture. Prigogine is best known for extending the second law of thermodynamics to systems that are far from equilibrium, and implying that new forms of ordered structures could exist under such disequilibrium conditions. Prigogine called these forms "dissipative structures," pointing out that they cannot exist independently of their environment. These dissipative structures have since been used to describe many phenomena of biology and ecology. Nicolis and Prigogine (1977) showed that the formation of dissipative structures allows order to be created from disorder in nonequilibrium systems.

The ecological core was pointed out in other books and papers (e.g., Capra, 1996; Ulanowicz, 1997). It was shown that the simplest ecosystems can display dynamical behavior table in certain situations. In certain cases, when riddled basins are found, even qualitative predictability is denied.

Thermodynamics is used increasingly in ecology to understand integral properties of ecosystems because it is a basic science that describes energy transformation from a holistic

Complexity and Complex Ecological Systems. https://doi.org/10.1016/B978-0-443-19237-1.00008-9

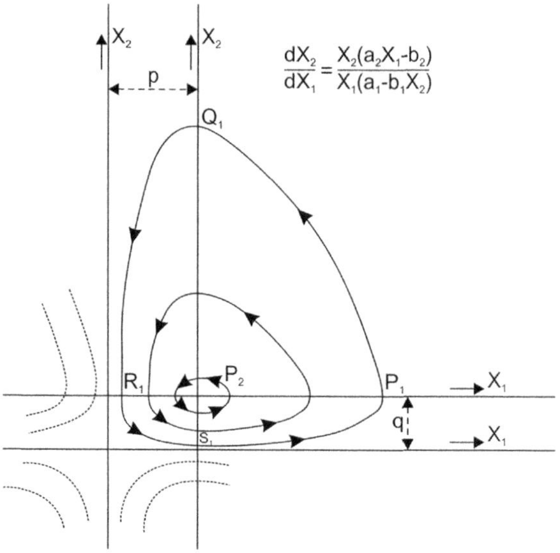

Fig. 1.1
Lotka-Volterra model is a classical scheme for interactions between populations.

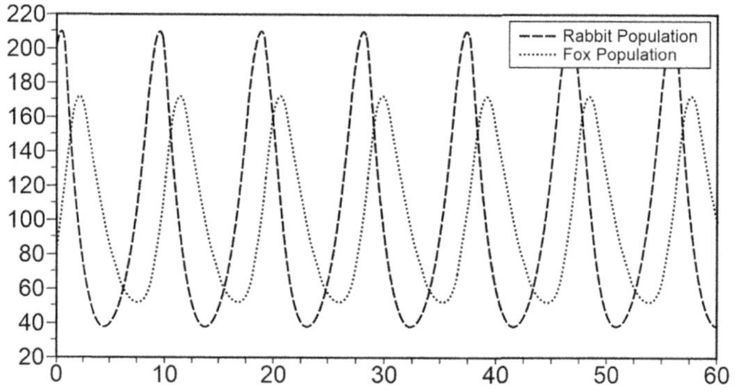

Fig. 1.2
A qualitative picture of the time variability of populations with predators (foxes) and preys (rabbits).

viewpoint. An ecosystem consists of interdependent living organisms that are also interdependent with their environment, all of which are involved in a continual transfer of energy and mass within a general state of equilibrium or disequilibrium. Thermodynamics can quantify in an exact way how "organized" or "disorganized" a system is, which is useful information to know when trying to understand how a dynamic ecosystem is behaving. The book by Faber et al. (1996) shows economists how to put the theory to use when trying to quantify the movement of goods and services through a sort of complex living system—a human society. This book may be supplemented by a volume on concepts and methods in ecological economics (Faber et al., 1996).

Schneider and Kay (1994a, b) believe that their thermodynamic paradigm facilitates the study of ecosystems developed from descriptive science to predictive science based on general macroscopic physics. Schrödinger (1944) points out that, at first glance, living systems seem to defy the second law of thermodynamics as it insists that, within closed systems, entropy should be maximized and disorder should reign. Living systems, however, are the antithesis of such disorder. They display excellent levels of order created from disorder. For instance, plants are highly ordered structures, which are synthesized from disordered atoms and molecules found in atmospheric gases and soils. By turning to nonequilibrium thermodynamics, Schrödinger recognizes that living systems exist well in a world of energy and material and flows (fluxes).

An organism stays alive in its highly organized state by taking energy from outside, from a larger encompassing system, and processing it to produce, within itself, a lower entropy, a more organized state. Schrödinger was first to recognize that life constitutes a far-from-equilibrium system that maintains its local level of organization at the expense of the larger global entropy budget. He proposes that to study living systems from a nonequilibrium perspective would reconcile biological self-organization and thermodynamics. Furthermore, he expects such a study to yield new principles of physics.

The paper by Schneider and Kay (1994a, b) takes on the task proposed by Schrödinger and expands on his thermodynamic view of life. Schneider and Kay explain that the second law of thermodynamics is not an impediment to the understanding of life but is necessary for a complete description of living processes. They further expand thermodynamics into the causality of living process and assert that the second law is a necessary but not a sufficient cause for life itself. In short, reexamination of thermodynamics shows that the second law underlies and determines the direction of many of the processes observed in the development of living systems.

The work of Schneider and Kay (1994a, b) links physics and biology at the macroscopic level and shows that biology is not an exception to physics, thus implying that to date we have probably misunderstood action of some rules of physics in biological context.

Regardless of the opinion of the reader concerning such reasoning, it would be difficult to overlook a fresh approach to thermodynamics in the research of Schneider and Kay, 1994a, b, 1995).

Since the time of Boltzmann and Gibbs, there have been major advances in thermodynamics, especially by Caratheodory, Hatsopoulos and Keenan, Kestin, Jaynes, and Tribus. Schneider and Kay (1994a, b) assumed the restated laws of thermodynamics of Hatsopoulos and Keenan (1965a, b) and Kestin (1993) and extended them so that disequilibrium processes and systems can be described in terms of gradients maintaining systems away from equilibrium. In their formulation, the second law states that as systems are moved away from equilibrium, they will use all available avenues to resist externally applied gradients.

Schneider and Kay's (1994a, b) formulation of the second law directly applies to complex systems with disequilibrium situations unlike classical statements which are limited to equilibrium or near equilibrium conditions. Away from equilibrium, highly ordered stable complex systems can emerge, develop, and grow at the expense of more disorder at higher levels in the system's hierarchy. These systems show the relevance and utility of these restatements of the second law by considering the classic example of dissipative structures, Benard cells. Their behavior proves that this paradigm can be applied to both physical and chemical systems and that it allows for a thermodynamically consistent explanation of the development of far-from-equilibrium complex systems including life.

As a case study, Schneider and Kay (1994a, b) select applications of their thermodynamic principles in the science of ecology. They view ecosystems as open thermodynamic systems with a large gradient impressed on them by the sun. The thermodynamic imperative of the restated second law is that these systems strive to reduce this gradient by action of all available physical and chemical processes. Consequently, ecosystems develop structures and functions selected to most effectively degrade or "dissipate" the imposed gradients while allowing for the continued existence of the ecosystem. In particular, they show that the unstressed ecosystem has structural and functional attributes that lead to more effective degradation of the energy provided to the ecosystem. Patterns of ecosystem growth, cycling, trophic structure, and efficiencies are explained by this paradigm.

A rigorous test of Schneider and Kay's hypothesis is the measurement of reradiated temperatures from terrestrial ecosystems. Schneider and Kay (1994a, b) argue that more mature ecosystems degrade incoming solar radiation into lower quality exergy at lower reradiated temperatures. Their data and arguments prove that not only more mature ecosystems are better degraders of energy (cooler final effect), but that airborne infrared thermal measurements of terrestrial ecosystems may offer a major breakthrough in providing measures of ecosystem health or integrity. Sieniutycz (1984) develops a variational approach to Brownian and molecular diffusion described by wave equations. The inertial terms of these equations cause some strange phenomena in the diffusional movement that enable molecules to enter the regions unpredicted by the classical Fick law.

1.2 Classical thermodynamics and the second law

As Schneider and Kay's basic ideas are built on the principles of original thermodynamics, their work starts with a brief discussion of classical thermodynamics. They believe that their approach to the theoretical issues of nonequilibrium thermodynamics is original, and this is also the opinion of the present author. In fact, their approach permits a more satisfactory discussion of the observed phenomena than the classical nonequilibrium thermodynamics.

The second law dictates that if there are any physical or chemical processes underway in a system, then the overall quality of the energy in that system will degrade, although the total quantity of energy in a closed system will remain unchanged. The second law can be stated as: any real process can only proceed in a direction that ensures a nonnegative entropy source (De Groot and Mazur, 1984). It is the property of the nonnegativeness of entropy source rather than entropy itself which draws Eddington's "arrow of time" in nature and defines the extent to which nature becomes more disordered or random (Schneider and Kay, 1994a, b). For other formulations and discussion of the significance of so-called natural variables, see Callen (1988).

All natural processes can be viewed in the light of the second law, and in all cases, this one-sided aspect of nature is observed. Schneider and Kay (1994a, b) recall some elementary examples, e.g., spontaneous flow of heat from a hotter reservoir to a colder reservoir until there is no longer a temperature difference or gradient. Another example is gas flow from a high pressure to a low pressure until there is no longer a pressure difference or gradient. If hot and cold water are mixed, the mixture comes to a uniform temperature. The resulting lukewarm water will not spontaneously unmix itself into hot and cold portions. Boltzmann would have restated the above example as: it is highly improbable that water will spontaneously separate into hot and cold portions, but it is not impossible (Schneider and Kay, 1994a, b). Boltzmann recasts thermodynamics in terms of energy microstates of matter. In this context, entropy reflects the number of different ways microstates can be combined to yield a particular macrostate. The larger the number of microstates for a given macrostate, the larger the entropy.

Consider, for example, a 10 compartment box with 10,000 marbles in one of the 10 compartments and the rest of the compartments being empty (Schneider and Kay, 1994a, b). If doors are opened between the compartments and the box is subjected to a pattern of random shaking, one would expect, over time, to see a distribution of about 1000 marbles per compartment, the distribution which has the largest number of possible microstates. This randomization of the marbles to the equiprobable distribution corresponds to the macrostate with the maximum entropy for the closed system. If one continued the shaking, it would be highly improbable but not impossible for all the marbles to reseparate themselves into the low-entropy configuration with 10,000 marbles in one compartment. The same logic is applied by Boltzmann to explain the macroscopic phenomena of thermodynamics in terms of microstates of matter. Systems tend to be the macrostate which has the largest number of corresponding accessible microstates.

1.3 Extended laws of thermodynamics

In 1908, thermodynamics was moved a step forward by Caratheodory's work (Caratheodory, 1976) where he developed a proof that the law of "entropy increase" is not the generalization of the second law. The more encompassing statement of the second law of thermodynamics is that "In the neighborhood of any given state of any closed system, there exist states which are inaccessible from it along any adiabatic path, reversible or irreversible." This statement of the second law, unlike the earlier formulations, does not depend on the contemporary definitions of entropy or temperature and applies equally well in the positive and negative temperature regimes.

Also, as early as in 1965, Hatsopoulos and Keenan (1965a, b) and Kestin (1993) put forward a principle which subsumes the 0th, 1st, and 2nd laws: "When an isolated system performs a process after the removal of a series of internal constraints, it will reach a unique state of equilibrium: this state of equilibrium is independent of the order in which the constraints are removed."

This statement has been called the Law of Stable Equilibrium by Hatsopoulos and Keenan and the Unified Principle of Thermodynamics by Kestin (Schneider and Kay, 1994a, b). The importance of the statement is that, unlike all earlier assertions that all real processes are irreversible, it dictates a direction and an end state for all real processes. All previous formulations of the second law tell us what systems cannot do (Schneider and Kay, 1994a, b). The present statement tells us what systems will do. An example of this phenomenon is two flasks connected with a closed stopcock (Schneider and Kay, 1994a, b). One flask holds 10,000 molecules of a gas. Upon removing the constraint (opening the stopcock), the system will come to its equilibrium state of 5000 molecules in each flask, with no gradient between the flasks.

To deal with the thermodynamics of nonequilibrium systems, Schneider and Kay (1994a, b) propose the following corollary that follows from the proof by Kestin of the Unified Principle of Thermodynamics; his proof shows that a system's equilibrium state is stable in the Lyapunov sense. Implicit in this "concluding corollary" is that a system will resist being removed from the equilibrium state. The degree to which a system has been moved from equilibrium is measured by gradients imposed on the system. According to Schneider and Kay (1994a, b), the thermodynamic principle which governs the behavior of systems is that, as they are moved away from equilibrium, they will utilize all avenues available to counter the applied gradients. As applied gradients increase, so does the system's ability to oppose further movement from equilibrium. In the discussion that follows Schneider and Kay (1994a, b), we shall refer to the above statement of the thermodynamic principle as the "restated second law." (The pre-Caratheodory statements, i.e., limited-value claims that entropy should increase, are referred to as the classical second law.)

A simple example of the relevant phenomenon is the Benard cell (Chandrasekhar, 1961). The experimental apparatus for studying the Benard cell consists of a highly instrumented insulated container enclosing a fluid. The bottom of the container is a heat source and the top is a cold reservoir. When the fluid is heated from below, it resists the applied gradients (ΔT) by dissipating heat through conduction. As the gradient is increased, the fluid develops convection cells. These structures increase the rate of dissipation. They are called Benard cells (Chandrasekhar, 1961).

There are many graphs in the literature illustrating instability phenomena of this sort. For an example, the reader is referred to Fig. 2c in Schneider and Kay (1994a, b) that shows a plot of a gradient (represented by the Rayleigh number, *Ra*), which is proportional to the temperature gradient ΔT against the available work expended in maintaining this gradient. The dynamics of the system are such that it becomes more and more difficult to move the system away from equilibrium. Proportionally, more available work must be spent for each incremental increase in gradient as the system gets further from equilibrium, i.e., ΔT increases. In chemical systems, LeChatelier's principle is another example of the restated second law. In his lectures on thermodynamics, Fermi (1956) points out that the effect of a change in external conditions on the equilibrium of a chemical reaction is prescribed by LeChatelier's principle. "If the external conditions of a thermodynamic system are altered, the equilibrium of the system will tend to move in such a direction as to oppose the change in the external conditions." Fermi (1956) stresses that if a chemical reaction were exothermal (i.e., A + B = C + D + heat), an increase in temperature will shift the chemical equilibrium to the left-hand side. Since the reaction from left to right is exothermal, the displacement of the equilibrium toward the left results in the absorption of heat and opposes the rise in temperature. Similarly, a change in pressure (at a constant temperature) results in a shift in the chemical equilibrium of reactions, which opposes the pressure change. Thermodynamic systems exhibiting temperature, pressure, and chemical equilibrium resist movement away from these equilibrium states. When moved away from their local equilibrium state, they shift their state in a way which opposes the applied gradients and moves the system back toward its local equilibrium attractor. The stronger the applied gradient, the greater the effect of the equilibrium attractor on the system. The reason that the restatement of the second law is a significant step forward for thermodynamics is that it tells us how systems will behave as they are moved away from equilibrium. Implicit in this is that this principle is applicable to nonequilibrium systems, something which is not true for classical formulations of the second law (Schneider and Kay, 1994a, b). In fact, their "restated second law" avoids the problems associated with state variables such as entropy which are only defined for equilibrium. Schneider and Kay's restatement of the second law sidesteps the problems of defining entropy and entropy production in nonequilibrium systems, an issue that has troubled researchers in nonequilibrium thermodynamics for years. By focusing on gradient destruction, Schneider and Kay avoid the problems encountered in works of Prigogine and others by using the entropy in which nonequilibrium systems can be described by their forces and requisite flows by using the well-developed methods of network thermodynamics (Katchalsky and Curran, 1965; Mikulecky, 2001; Peusner, 1986; Swenson, 1989).

The spontaneous organization of water due to convection, once convection begins and the dissipative structure forms a pattern of hexagonal Benard cells appear, was illustrated in the literature a number of times (compare with Schneider and Kay, 1994a, b, his Fig. 1; Capra, 1996, Fig. 5.1, p. 87, etc.)

1.4 Dissipative structures, degraders, and related problems

Prigogine and his co-workers have shown that dissipative structures self-organize through fluctuations, small instabilities which lead to irreversible bifurcations and new stable system states (Glansdorff and Prigogine, 1971; Nicolis and Prigogine, 1977, 1979). In this sense, the future states of such systems are regarded as not deterministic. Dissipative structures are stable over a finite range of conditions and are sensitive to fluxes and flows from outside the system. Glansdorff and Prigogine (1971) show that these thermodynamic relationships are best represented by coupled nonlinear relationships, i.e., autocatalytic positive feedback cycles, many of which lead to stable macroscopic structures which exist away from the equilibrium state. Convection cells, hurricanes, autocatalytic chemical reactions, and living systems are all examples of far-from-equilibrium dissipative structures, which exhibit coherent behavior. The transition between conduction and convection in a heated fluid is a striking example of emergent coherent organization in response to an external energy input. A thorough analysis of these simple physical systems allows to develop a number of general thermodynamic principles applicable to the development of complex systems as they emerge at some distance away from equilibrium (Schneider and Kay, 1994a, b).

Benard and Rayleigh (Chandrasekhar, 1961; Schneider and Kay, 1994a, b) conducted carefully designed experiments to study this transition. The lower surface of the experimental apparatus is heated, and the upper surface is kept at a cooler temperature; see, e.g., Fig. 1 in Schneider and Kay (1994a, b). Hence, a temperature gradient is induced across the fluid. The initial heat flow through the system is controlled by conduction; the energy is then transferred by molecular interaction. When the heat flux reaches a critical value of the temperature gradient, the system becomes unstable and convective overturning emerges. The convective structures appear in highly structured coherent hexagonal surface patterns (Benard cells) in the fluids. The coherent kinetic structuring increases the rate of heat transfer and gradient destruction in the system.

As shown by Schneider and Kay (1994a, b), the transition from noncoherent, molecule to molecule, heat transfer to a coherent structure results in excess of 10 molecules acting together in an organized manner. This seemingly improbable occurrence is the direct result of the applied temperature gradient and is the system's response to attempts to move it away from equilibrium. Schneider and Kay (1994a, b) studied the Benard cell phenomena in detail, exploiting the original data sets collected by Silveston (1957) and Brown (1973). Their analyses are significant, in that they calculated for the first time the entropy production, exergy drop, and available work destruction, resulting from these organizing events (see Fig. 2 in Schneider and Kay, 1994a, b). This analysis shows that as the gradient increases, it becomes harder to increase this gradient. Fig. 2 in Schneider and Kay (1994a, b) shows (A) the heat dissipation rate W(J/s) and (B) entropy production rate *P* (J/s/K) versus Rayleigh number *Ra* as nondimensional measure of the temperature gradient. The sketch is based on experimental data of Silveston (1957) and Brown (1973).

Initially, the temperature gradient in the apparatus is accommodated solely by random conductive activity. When the gradient is raised, a combination of factors including surface tension effects and gravitational fluid instability converts the system to a mixed conductive-convective heat transfer system. The transition to coherent behavior occurs at *Ra* 51,760. See Table 1 and Figs. 1 and 2 in Schneider and Kay (1994a, b), which illustrate behavior of heat dissipation rate (W) and power as important part of these results.

As shown in Fig. 2 of Schneider and Kay (1994a, b), with the onset of convection, there is a dramatic increase in the heat transfer rate across the system. From the literature, especially Chandrasekhar (1961), and also from Schneiders and Kay's comprehensive analysis of the Brown's and Silveston's data, Brown (1973), and Silveston (1957), we can observe the following behavior for these systems:

1. Heat dissipation rate (transfer of heat between the plates, W) is a linearly increasing function of the gradient ΔT (see Fig. 2a in Schneider and Kay (1994a, b), recalling that *Ra* is proportional to ΔT).
2. Entropy production rate, P, vs ΔT increases in a nonlinear way (Fig. 2b therein).
3. Destruction rate of the classical exergy Θ vs. gradient ΔT increases in a nonlinear way, the shape of the curve being the same as P vs. ΔT.
4. Points 2 and 3 imply that with the gradient increase, it is harder (requires more available work) to incrementally increase the gradient. The further from equilibrium the system is, the more it resists being moved further from equilibrium. In any real system, there is an upper limit to the gradient that can be applied to the system.
5. Once convection occurs, the temperature profile within the fluid is vertically isothermal outside the boundary layer, i.e., the temperature in the convection cells is constant, thus effectively removing the gradient through most of the fluid.
6. As the gradient increases, further critical points are reached. At each critical point, the boundary layer depth decreases.
7. Point 1 is valid because the rate of heat transfer is controlled by the rate of heat flow across the boundary layer, i.e., by conduction which is a linear process. This process is also responsible for most of the entropy production, as there will be little production due to convection. The slope change is caused by the decrease in the boundary layer depth at a mode change (critical point). (Recall that $Q = k\Delta T/l$, thus as l decreases, slope k/l increases.)
8. Nusselt number, Nu, equals $Q/Q_c = P/P_C = \Theta/\Theta_C$, that is, in Benard cells, the increase in dissipation at any point due to the emergent process is equal to the increase in degradation at any point due to the emergent process. This is true for any process that involves only heat transfer. Otherwise, degradation differs from dissipation. (As noted by Schneider and Kay (1994a, b), Prigogine at times mistakenly used these terms interchangeably.)

9. The principle governing these systems is not one of maximum entropy production, but rather one of entropy production change being positive semidefinite as you increase the gradient. See Point 7 above. The interesting question is how much structure emerges for a given gradient, and how much resistance exists to increasing the gradient further.

As the temperature difference increases, there are a number of further transitions at which the hexagonal cells reorganize themselves so that the cost of increasing the temperature gradient escalates even more quickly. Ultimately, the system becomes chaotic and the dissipation is maximum in this regime. The point of this example is that in a simple physical system, new structures emerge which better resist the application of an external gradient. The Benard cell phenomenon is an excellent example of the nonequilibrium restated second law (Schneider and Kay, 1994a, b).

Other physical, chemical, and living systems exhibit similar behavior. The more a system is moved from equilibrium, the more sophisticated are its mechanisms for resisting being moved from equilibrium. This behavior is not sensible from a classical second law perspective, but this behavior is predicted by the restated second law. No longer is the emergence of coherent self-organizing structures a surprise, but rather it is an expected response of a system as it attempts to resist and degrade externally applied gradients, which would move the system away from equilibrium. See Schneider and Kay (1994a, b) for a more comprehensive discussion which, in particular, mimics meterological phenomena.

Until now, we have focused the discussion on simple physical systems and answering the question how thermodynamic gradients drive self-organization. The literature is replete with similar phenomena in dynamic chemical systems. Prigogine and many other representatives of the Brussels School (Glansdorff and Prigogine, 1971; Nicolis and Prigogine, 1977, 1979) have documented the thermodynamic behavior of these chemical reaction systems. Chemical gradients arise in dissipative autocatalytic reactions, examples of which are found in simple inorganic chemical systems, in protein synthesis reactions, or phosphorylation, polymerization, and hydrolytic autocatalytic reactions (Schneider and Kay, 1994a, b).

1.5 Destructive entropy production

On the other hand, Mansson and Lindgren (1990) stress the role of the transformed information for creation processes by showing how destructive processes of entropy production are related to creative processes of formation. Nonlinear molecular mechanism gives rise to critical points, instabilities, bifurcations, chaotic behavior, and oscillations. This leads, in general, to various forms of organization (Ebeling, 1985; Ebeling et al., 1986; Ebeling and Feistel, 1992; Ebeling and Klimontovich, 1984; Nicolis and Prigogine, 1977; Zainetdinov, 1999).

Systems of autocatalytic reactions are a form of positive feedback, where the activity of the system or reaction augments itself in the form of self-reinforcing reactions. An example is a reaction where a compound A catalyzes the formation of compound B and B accelerates the formation of A; then, the overall set of reactions is an autocatalytic or positive feedback cycle.

Ulanowicz (1986) notes that, in autocatalysis, activity of any element in the cycle engenders greater activity in all the other elements, thus stimulating an aggregate activity of the whole cycle. Such self-reinforcing catalytic activity is self-organizing and is an important way of increasing the dissipative capacity of the system. Cycling and autocatalysis are fundamental in nonequilibrium systems (Schneider and Kay, 1994a, b).

The notion of dissipative systems as gradient dissipators holds for disequilibrium physical and chemical systems and describes processes of emergence and development of complex systems (Schneider and Kay, 1994a, b). Not only are the processes of these dissipative systems consistent with the restated second law, it is expected that they exist wherever there are gradients.

Klimontovich (1986, 1991) compares entropies of laminar and turbulent motions with respect to the velocity of a laminar flow and the velocity of an averaged turbulent flow, for the same values of the average kinetic energies. He shows that when transition occurs from laminar flow to turbulent flow, the entropy production and entropy itself decrease. This proves that the disequilibrium phase transition from laminar flow to turbulent flow transfers the system to a more ordered state. He also proves that the turbulent motion should be regarded as motion having a lower temperature than laminar motion. Klimontovich's (1999) self-organization treatment using his "norm of chaoticity" is carried out on a number (classical and quantum) of examples of physical systems including also an example of a medical-biological system.

Applying his turbulence theory (Klimontovich, 1982), Klimontovich (1991) examines the link between the turbulent motion and the structure of chaos. Klimontovich (1999) considers entropy, information, and some relative criteria of order for states of open systems. He points out that two meanings of the word "information" are known in the theory of communication. The first meaning coincides in form with Boltzmann's entropy. The second refers to difference between unconditional and conditional entropies. Two kinds of open systems are considered. Systems of the first kind, with zero value of a controlling parameter C, are in equilibrium. Systems of the second kind, with a finite value of C, are in disequilibrium. In self-organization, associated with escaping from equilibrium, information is increased. For open systems and all values of C, the conservation law holds for the sum of information and entropy with all values of the controlling parameter. Zainetdinov (1999) attempts to attribute informational negentropy to self-organization processes in open systems.

1.6 The origin of life: A brief introduction

The origin of prebiotic life may be regarded as the development of another route for the dissipation of induced energy gradients. Life with its requisite ability to reproduce insures that these dissipative pathways continue, having evolved strategies to maintain the dissipative structures in the face of a fluctuating physical environment. Schneider and Kay (1994a, b) suggest that living systems are dynamic dissipative systems with encoded memories, the gene with its DNA, which allow dissipative processes to continue without having to restart the dissipative process via stochastic events. They see living systems as sophisticated minitornados, with a memory (its DNA), whose Aristotelian "final cause" may be the second law of thermodynamics. However, the role of thermodynamics in living processes should not be exaggerated. In their summarizing considerations on the origin of life, Schneider and Kay (1994a, b) maintain that "the restated second law is a necessary but not a sufficient condition for life," and, in addition, they state: "Life should be viewed as the most sophisticated (until now) end in the continuum of development of natural dissipative processes from physical to chemical to autocatalytic to living systems." Life should not be seen as an isolated event. Rather, it is the emergence of yet another class of processes aimed at the degradation/consumption of energy attributed to thermodynamic gradients. Life should be viewed as the most sophisticated (until now) end in the continuum of development attributed to natural dissipative structures from physical to chemical to autocatalytic to living systems.

Schneider and Kay (1994a, b) believe that autocatalytic chemical reactions are the backbone of chemical systems leading to origins of life. They regard papers of Eigen (1971) and Eigen and Schuster (1979) as contributions that rightly link autocatalytic and self-reproductive macromolecular species with thermodynamic vision of the origin of life. Ishida (1981) summarizes the essence of Eigen's work (Eigen, 1971; Eigen and Schuster, 1979).

Occasionally, the second law is stated as follows: any real process can only proceed in a direction which results in an entropy increase, but this statement is in general incorrect (Schneider and Kay, 1995). The correct statement reads: any real process can only proceed in the direction which ensures a nonnegative entropy source. This latter formulation is consistent with Caratheodory's 1908 formulation of the second law. In 1908, thermodynamics progressed through the work of Caratheodory (Kestin, 1993) when he developed a proof which shows that the law of "entropy increase" is not the general statement of the second law. The more encompassing statement of the second law of thermodynamics ensures that "In the neighborhood of any given state of any closed system, there exist states which are inaccessible from it along any adiabatic path reversible or irreversible." Unlike in earlier statements, this does not depend on the nature of the system, nor on concepts of entropy or temperature.

Hatsopoulos and Keenan (1965) and Kestin (1968) subsumed the 0th, 1st, and 2nd laws into a Unified Principle of Thermodynamics: "When an isolated system performs a process after the removal of a series of internal constraints, it will reach a unique state of equilibrium: this state of equilibrium is independent of the order in which the constraints are removed." This describes systems of some distance from equilibrium, but not constrained to be in a disequilibrium state. This also dictates direction and an end state for real processes and says that a system will come to a local equilibrium as constraints permit (Schneider and Kay, 1995).

Life can be viewed as a far-from-equilibrium dissipative structure that maintains its local level of organization at the expense of producing entropy in the environment (Schneider and Kay, 1995). If we treat the Earth as an open thermodynamic system with a large gradient impressed on it by the Sun, the restated second law implies that the system should reduce this gradient by using all physical and chemical processes available to it. Schneider and Kay (1994a, b, 1995) ensure that life exists on Earth to dissipate the solar-induced gradient, and as such, life is a manifestation of the restated second law. However, in a common but different language, one can say that life exists because our terrestrial system is capable of absorbing solar energy, which induces birth processes. Living systems are far-from-equilibrium dissipative structures that have sophisticated potential for reducing radiation gradients and inducing birth processes on Earth (Kay, 1984; Ulanowicz and Hannon, 1987).

Consequently, the origin of life means the development of an alternative road for the dissipation of induced energy gradients (absorption of energy at flow). Life ensures that its dissipative pathways continually evolve strategies to maintain it in the fluctuating physical environment. With genes as encoded memories, living systems are dynamic dissipative structures which allow to continue life as a response to the thermodynamic imperative of dissipating gradients (Kay, 1984; Schneider, 1988). Biological growth occurs when the system adds more of the same types of pathways for degrading imposed gradients. Biological development takes place when new types of pathways for degrading imposed gradients emerge in the system.

Plant growth—typical in ecology—is an attempt to capture solar energy and dissipate usable gradients (or related driving fluxes). Plants of many species arrange themselves into assemblies to increase leaf area so as to maximize driving energy capture and its degradation. The gross energy budgets of terrestrial plants show that the majority of their energy use is for evapotranspiration. The more energy available to be distributed among species, the more pathways are available for total energy degradation. Trophic levels and food chains are based upon photosynthetic fixed material and further dissipate these gradients by making more highly ordered structures. Thus, we would expect greater diversity of species to occur where there is more available energy. Species diversity and trophic levels are vastly greater at the equator, where 5/6 of the Earth's solar radiation occurs, and there is more of a gradient to reduce (Schneider and Kay, 1995).

1.7 Thermodynamics of ecosystems as energy degraders

Ecosystems are biotic, physical, and chemical components of nature acting together as disequilibrium dissipative processes. Ecosystem development should increase energy degradation whenever it is obedient to the restated second law (Schneider and Kay, 1995). This hypothesis can be tested by observing the energetics of ecosystem development during the successional process or when ecosystems are stressed. As ecosystems develop or become mature, they should increase their total dissipation and should develop more complex structures with greater diversity and more hierarchical levels to assist in energy degradation (Schneider, 1988; Kay and Schneider, 1992a, b). Successful species are those that funnel more energy into their own production and reproduction and contribute to autocatalytic processes, thereby increasing the total dissipation in the ecosystem.

Lotka (1922) and Odum and Pinkerton (1955) have suggested that biological systems that survive are those that develop the most power in flow and use this power to the best satisfaction of their needs for survival. Schneider and Kay (1995) believe that a better alternative to these "power laws" may be that biosystems develop so as to increase their energy degradation rate and that biological growth, ecosystem development, and evolution represent the rise of new dissipative pathways. In other words, ecosystems develop in a way that increases the amount of exergy they capture and utilize. As a consequence, as ecosystems develop, exergy of the outgoing energy decreases. It is in this sense that ecosystems develop the most power, that is, they make the most effective use of the exergy in the incoming energy, while at the same time increasing the amount of energy they capture.

Ecologists have developed analytical methods that allow analysis of material-energy flows through ecosystems (Kay et al., 1989). With these methods, it is possible to determine the energy flow and assess how the energy is partitioned in the ecosystem. Schneider and Kay (1995) analyze data for carbon-energy flows in two aquatic tidal marsh ecosystems adjacent to a large power generating facility on the Crystal River in Florida (Ulanowicz, 1986). The ecosystems considered are a "stressed" and a "control" systems. The "stressed" ecosystem is exposed to hot water effluent from the nuclear power station. The "control" ecosystem is not exposed to the effluent but is otherwise exposed to the same environmental conditions. All the flows dropped in the stressed ecosystem. It follows that the stress has resulted in the ecosystem shrinking in size, in terms of biomass, its consumption of resources, in material and energy cycling and its ability to degrade and dissipate incoming energy Schneider and Kay (1995).

The impact of the effluent from the power station heating water has been to decrease the size of the "stressed" ecosystem and its consumption of resources while impacting on its ability to retain the resources it has captured. This analysis suggests that the function and structure of ecosystems follow the development path predicted by the behavior of nonequilibrium thermodynamic structures and the application of these behaviors to ecosystem development patterns (Schneider and Kay, 1995). The energetics of terrestrial ecosystems provides another test of the thesis that ecosystems develop so as to degrade energy more effectively. It turns out that more developed dissipative structures degrade more energy. Thus, we expect a more mature ecosystem to degrade the exergy content of the energy it captures more completely than a less developed ecosystem.

Luvall and Holbo (1989, 1991) measured surface temperatures of various ecosystems using a thermal infrared multispectral scanner (TIMS). Their data show a solid trend: when other variables are constant, the more developed the ecosystem, the colder is its surface temperature and the more the degradation of its reradiated energy. The data analysis shows that ecosystems develop structure and function that degrades imposed energy gradients more and more effectively (Schneider and Kay, 1994a, b, Schneider and Kay, 1995). Their study of energetics treats ecosystems as open systems with high-quality energy pumped into them. Clearly, an open system with high-quality energy pumped into it can be moved away from equilibrium. But nature resists movement away from equilibrium. So ecosystems, as open systems, respond, whenever possible, with the spontaneous emergence of organized behavior that consumes the high-quality energy in building and maintaining the newly emerged structure. This decreases the ability of the high-quality energy to move the system further away from equilibrium. Such self-organization is characterized by abrupt changes that occur as a new set of interactions and activities by components and the whole system emerges. This emergence of organized behavior, the essence of life, is now understood and predicted by thermodynamics. As more high-quality energy is pumped into an ecosystem, more organization emerges to dissipate (degrade) the energy. This way, Schneider and Kay (1995) find order emerging from disorder in the situation causing even more disorder.

1.8 Order from disorder and order from order

Complex systems stretch from ordinary complexity (Prigoginean systems, tornadoes, Benard cells, autocatalytic reaction systems) to emergent complexity perhaps including human socioeconomics systems. Living systems are at the more sophisticated end of the continuum: they must function within the system and environment they are part of. If a living system does not respect the circumstances of the supersystem it is part of, it will be selected against this supersystem (Schneider and Kay, 1995). The supersystem imposes a set of constraints on the behavior of the system and those living systems that are evolutionarily successful and are prepared to live within them.

When a new living system is generated after the death of an earlier one, it would make the self-organization process more efficient if it is constrained to variations that have a high probability of success. Genes play this role in constraining the self-organization process to those options that have a high probability of success. They are a record of successful self-organization. Genes are not the mechanism of development; the mechanism is self-organization. Genes bound and constrain the process of self-organization. At higher hierarchical levels, other devices constrain self-organization. The ability of an ecosystem to regenerate is a function of the species available for the regeneration process (Schneider and Kay, 1995).

Given that living systems go through a constant cycle of many processes (birth/development/regeneration/death), preserving information about what works and what does not is crucial for the continuation of life (Kay, 1984). This is the role of the gene and, at a larger scale, biodiversity to act as information data bases about self-organization strategies that work. This is the connection between the order from order and order from disorder themes of Schrödinger. Life emerges because thermodynamics mandates order from disorder whenever sufficient thermodynamic and environmental conditions exist. But if life is to continue, the same rules require that it is able to regenerate, that is to create order from order. Life cannot exist without both processes, order from disorder to generate life and order from order to ensure the continuance of life (Schneider and Kay, 1995). Life represents a balance between the imperatives of survival and energy degradation. We could add: degradation by consumption of energy and matter to sustain life.

A synthetic review of chemical stability properties has been presented by Farkas and Noszticzius (1992) for the three basic groups of the systems showing explosive, conservative, or damped behavior, depending on parameters of a kinetic model. Their unified treatment of chemical oscillators considers these oscillators as thermodynamic systems far from equilibrium. The researchers start with the theory of the Belousov-Zhabotinsky (B-Z) reaction pointing out the role of positive and negative feedback (Zhabotinsky, 1974).

Consequently, Farkas and Noszticzius (1992) introduce and analyze certain (two-dimensional) generalized Lotka-Volterra models. They show that these models can be conservative, dissipative, or explosive, depending on the value of a parameter. The transition from dissipative to explosive behavior occurs via a critical Hopf bifurcation. Their system is conservative at certain critical values of the parameter. The stability properties can be examined by testing the sign of a Lyapunov function selected as an integral of the conservative system. By adding a limiting reaction to an explosive network, a limit cycle oscillator surrounding an unstable singular point is obtained. Some general thermodynamic aspects are discussed pointing out the distinct role of oscillatory dynamics as an inherent far-from-equilibrium phenomenon. The existence of the attractors is recognized as a consequence of the second law. The paper of Farkas and Noszticzius (1992) gives a pedagogical perspective on the broad class of problems involving the time order in homogeneous chemical systems.

Broad exergy applications stretch from the realm of thermal technology to chemical, metallurgical, and other technologies (Szargut et al., 1993). Particularly fruitful are ecological applications of exergy for the analysis of cumulative exergy consumption and cumulative exergy losses (Szargut, 1990).

For considerations regarding the appropriateness of early optimization criteria and ecological performance functions, see: Angulo-Brown, 1991; Ayres, 1978, 1999; Ayres et al., 1996, 1998; Baumgärtner et al., 1996; Boltzmann, 1974; Boulding, 1978; Borys, 1999; Faber et al., 1995, 1996; Jørgensen, 1988, 1997; Kay and Schneider, 1992a,b; Odum, 1971, 1988; Szargut, 1986, 1990; Szargut et al., 1993; Valero, 1995, 1996; Yan, 1993, and some others.

References

Angulo-Brown, F., 1991. An ecological optimization-criterion for finite-time heat engines. J. Appl. Phys. 69, 7465–7469.

Ayres, R.U., 1978. Resources Environment and Economics—Applications of the Materials/Energy Balance Principle. Wiley, New York.

Ayres, R.U., 1999. The second law, the fourth law, recycling, and limits to growth. Ecol. Econ. 29, 473–483.

Ayres, R.U., Ayres, L.W., Martinas, K., 1996. Eco-Thermodynamics: Exergy and Life Cycle Analysis. Working Paper (96/./EPS), INSEAD, Fontainebleau, France.

Ayres, R.U., Ayres, L.W., Martinas, K., 1998. Exergy, waste accounting, and life-cycle analysis. Energy 23, 355–363.

Baumgärtner, S., Faber, M., Proops, R., 1996. The use of the entropy concept in ecological economics. In: Faber, M., Manstetten, R., Proops, J. (Eds.), Ecological Economics—Concepts and Methods. Edward Elgar, Cheltenham, pp. 115–135.

Boltzmann, L., 1974. The second law of thermodynamics. In: McGinness, B. (Ed.), Ludwig Boltzmann, Theoretical Physics and Philosophical Problems. D. Reidel Publishing Co, Dordrecht.

Borys, T., 1999. Indicators of Eco-Development (Wskaźniki Ekorozwoju). Ekon. S ́r. 22, 12–25. Białystok, Poland.

Boulding, K., 1978. Ecodynamics. Sage, Beverly Hills.

Callen, H., 1988. Thermodynamics and an Introduction to Thermostatistics. Wiley, New York.

Capra, F., 1996. The Web of Life: A New Scientific Understanding of Living Systems. Anchor Books, New York.

Caratheodory, C., 1976. Investigations into the foundations of thermodynamics. In: Kestin, J. (Ed.), The Second Law of Thermodynamics. Within Benchmark Papers on Energy. vol. 5. Dowden, Hutchinson, and Ross, Stroudsburg, PA, pp. 229–256.

Chandrasekhar, S., 1961. Hydrodynamic and Hydromagnetic Stability. Clarendon Press, Oxford.

De Groot, S.R., Mazur, P., 1984. Nonequilibrium Thermodynamics. Dover, New York.

Ebeling, W., 1985. Thermodynamics of self-organization and evolution. Biomed. Biochim. Acta 44, 831–838.

Ebeling, W., Feistel, R., 1992. Theory of self-organization and evolution: the role of entropy, value and information. J. Non-Equilib. Thermodyn. 17, 303–332.

Ebeling, W., Klimontovich, Y.L., 1984. Self-Organization and Turbulence in Liquids. Teubner Verlagsgeselshaft, Leipzig.

Ebeling, W., Engel-Herbert, H., Herzel, H., 1986. On the entropy of dissipative and turbulent structures. Ann. Phys. (Leipzig) 408 (3–5), 187–195.

Eigen, M., 1971. Self-organization of matter and the evolution of biological macro molecules. Naturwissenschaften 58 (10), 465–523.

Eigen, M., Schuster, P., 1979. The Hypercycle: A Principle of Natural Self-Organization. Springer Verlag, Berlin.

Faber, M., Jost, F., Manstetten, R., 1995. Limits and perspectives on the concept of sustainable development. Econ. Appl. 48, 233–251.

Faber, M., Manstetten, R., Proops, J., 1996. Ecological Economics—Concepts and Methods. Edward Elgar, Cheltenham, UK.

Farkas, H., Noszticzius, Z., 1992. Explosive, stable and oscillatory behavior in some chemical systems. In: Flow, Diffusion and Transport Processes. Advances in Thermodynamics Series. vol. 6. Taylor & Francis, New York, pp. 303–339.

Fermi, 1956. Thermodynamics. Dover, UK.

Glansdorff, P., Prigogine, I., 1971. Thermodynamic Theory of Structure Stability and Fluctuations. Wiley, New York.

Hatsopoulos, G., Keenan, J., 1965a. Principles of General Thermodynamics. John Wiley, New York.

Hatsopoulos, G., Keenan, J., 1965b. Principles of General Thermodynamics. John Wiley, New York.

Ishida, K., 1981. Non-equilibrium thermodynamics of the selection of biological macromolecules. J. Theor. Biol. 88, 257–273.

Jørgensen, S.E., 1988. Use of models as an experimental tool to show that structural changes are accompanied by increased exergy. Ecol. Model. 41, 117–126.

Jørgensen, S.E., 1997. Integration of Ecosystem Theories: A Pattern, second ed. Kluwer Academic Publishers, Dordrecht.

Katchalsky, A., Curran, P., 1965. Nonequilibrium Thermodynamics in Biophysics. Harward Univ. Press and and de Gruyter.

Kay, J.J., 1984. Self-Organization in Living Systems. Ph.D. Thesis, University of Waterloo. Systems Design Engineering, Waterloo, Ontario.

Kay, J., Schneider, E.D., 1992a. Thermodynamics and measures of ecological integrity. In: Proceedings of Ecological Indicators. Elsevier, Amsterdam, pp. 159–182.

Kay, J., Schneider, E., 1992b. Thermodynamics and measures of ecosystem integrity. In: McKenzie, D.H., Hyatt, D.E., McDonald, V.J. (Eds.), Ecological Indicators. vol. 1. Elsevier, New York, pp. 159–182.

Kay, J.J., Graham, L., Ulanowicz, R.E., 1989. A detailed guide to network analysis. In: Wulff, F., Field, J.G., Mann, K.H. (Eds.), Network Analysis in Marine Ecosystems. Coastal and Estuarine Studies. vol. 32. Springer-Verlag, New York. pp. 16–61.4.

Kestin, J., 1993. Internal variables in the local equilibrium approximation. J. Non-Equilib. Thermodyn. 18, 360–379.

Klimontovich, Y.L., 1982. Kinetic Theory of Nonideal Gases and Nonideal Plasmas. Pergamon, Oxford.

Klimontovich, Y.L., 1986. Statistical Physics. Harwood Academic Publishers, Chur.

Klimontovich, Y.L., 1991. Turbulent Motion and the Structure of Chaos. Kluwer Academics, Dordrecht.

Klimontovich, Y.L., 1999. Entropy, information and criteria of order in open systems. Nonlinear Phenom. Complex Syst. 2 (4), 1–25.

Lotka, A.J., 1922. Contribution to the energetics of evolution. Proc. Natl. Acad. Sci. U. S. A. 8, 147–151.

Luvall, J.C., Holbo, H.R., 1989. Measurements of short term thermal responses of coniferous forest canopies using thermal scanner data. Remote Sens. Environ. 27, 1–10.

Luvall, J.C., Holbo, H.R., 1991. Thermal remote sensing methods in landscape ecology. Ecol. Stud. 82, 127–152.

Mansson, B.A.G., Lindgren, K., 1990. Thermodynamics, information and structure. In: Nonequilibrium Theory and Extremum Principles. Advances in Thermodynamics Series. vol. 3. Taylor and Francis, New York, pp. 95–128.

Mikulecky, D.C., 2001. A simple network thermodynamic method for modelling series-parallel coupled flows II: the nonlinear theory with applications to coupled solute and volume flow in series membrane. J. Theor. Biol. 69, 511–541.

Nicolis, G., Prigogine, I., 1977. Self-Organization in Nonequilibrium Systems. From Dissipative Structures to Order through Fluctuations. John Wiley & Sons, New York.

Nicolis, G., Prigogine, I., 1979. Irreversible processes at nonequilibrium steady states and Lyapunov functions. Proc. Nat. Acad. Sci. USA 76 (12), 6060–6061.

Odum, 1971. Environment, Power and Society. Wiley, New York.

Odum, H.T., 1988. Self-organization, transformity, and information. Science 242, 1132–1139.

Odum, H.T., Pinkerton, R.C., 1955. Trials speed regulator: the optimum efficiency for maximum power output in physical and biological systems. Am. Sci. 43 (2), 331–343.

Peusner, L., 1986. Studies in Network Thermodynamics. Elsevier, Amsterdam.

Prigogine, I., 1961. Introduction to Thermodynamics of Irreversible Processes. Wiley, New York.

Schneider, E., 1988. Thermodynamics, information, and evolution: new perspectives on physical and biological evolution. In: Weber, B.H., Depew, D.J., Smith, J.D. (Eds.), Entropy, Information, and Evolution: New Perspectives on Physical and Biological Evolution. MIT Press, Boston, pp. 108–138.

Schneider, E.D., Kay, J.J., 1994a. Life as a manifestation of the second law of thermodynamics. Math. Comput. Model. 19 (6–8), 25–48.

Schneider, E., Kay, J., 1994b. Complexity and thermodynamics: towards a new ecology. Future 24, 626–647.

Schneider, E.D., Kay, J.J., 1995. Order from disorder: the thermodynamics of complexity in biology. In: Murphy, M.P., O'Neill, L.A.J. (Eds.), What Is Life: The Next Fifty Years. Reflections on the Future of Biology. Cambridge University Press, pp. 161–172.

Schrödinger, E., 1944. What Is Life? The Physical Aspect of the Living Cell. Cambridge University Press, Cambridge, UK.

Sieniutycz, S., 1984. The Variational approach to Brownian and molecular diffusion described by wave equations. Chem. Eng. Sci. 39, 71–80.

Swenson, R., 1989. Emergent attractors and the law of maximum entropy production. Foundations to the theory of general evolution. Syst. Res. 6 (3), 187–197.

Szargut, J., 1986. Application of exergy for the calculation of ecological cost. Bull. Pol. Acad. Sci.: Tech. Sci. 34 (1986), 475–480.

Szargut, J., 1990. Analysis of cumulative exergy consumption and cumulative exergy losses. In: Sieniutycz, S., Salamon, P. (Eds.), Finite-Time Thermodynamics and Thermoeconomics. Advances in Thermodynamics, vol. 4. Taylor and Francis, New York, pp. 278–302.

Szargut, J., Morris, D.R., Steward, F., 1988. Exergy Analysis of Thermal, Chemical and Metallurgical Processes. Hemisphere, New York.
Szargut, J., Kolenda, Z., Tsatsaronis, G., Ziębik, A. (Eds.), 1993. Energy systems and ecology. Proceedings of the International Conference Energy Systems and Ecology ENSEC'93, Cracow, Poland, July 5–9. vol. 2, pp. 339–350.
Ulanowicz, R.E., 1986. Growth and Development: Ecosystem Phenomenology. Springer-Verlag, New York.
Ulanowicz, R.E., 1997. Ecology, the Ascendent Perspective. Columbia University Press, New York.
Ulanowicz, R.E., Hannon, B.M., 1987. Life and the production of entropy. Proc. R. Soc. Lond. B 232, 181–192.
Valero, A., 1995. Thermoeconomics: the meeting point of thermodynamics, economics and ecology. In: Sciubba, E., Moran, M. (Eds.), Second Law Analysis of Energy Systems: Towards the 21st Century, Circus, Rome, pp. 293–305.
Valero, A., 1996. Thermoeconomics as a Conceptual Basis for Energy-Ecological Analysis. https://www.researchgate.net/publication/254337577.
Yan, Z., 1993. Comment on "ecological optimization criterion for finite-time heat engines". J. Appl. Phys. 73, 3583.
Zainetdinov, R.I., 1999. Dynamics of informational negentropy associated with self-organization process in open system. Chaos Soliton. Fract. 10, 1425–1435.

Further reading

Ahlborn, B.K., 1999. Thermodynamic limits of body dimension of warm blooded animals. J. Non-Equilib. Thermodyn. 40, 407–504.
Ahlborn, B.K., Blake, R.W., 1999. Lower size limit of aquatic mammals. Am. J. Phys. 67, 1–3.
Aoki, I., 1995. Entropy production in living systems: from organisms to ecosystems. Thermochim. Acta 250, 359–370.
Ayres, R.U., 1998. Eco-thermodynamics: economics and the second law. Ecol. Econ. 26, 189–209.
Kneese, A.V., Ayres, R.U., d'Arge, R.C., 1972. Economics and the Environment: A Materials Balance Approach. Resources for the Future, Washington.
Lotka, A.J., 1920. Undamped oscillations derived from the law of mass action. J. Am. Chem. Soc. 42, 1595–1599.
Lozano, M.A., Valero, A., Serra, L., 1993. Theory of exergetic cost and thermoeconomic optimization. In: Szargut, J., Kolenda, Z., Tsatsaronis, G., Ziębik, A. (Eds.), Proceedings of the International Conference Energy Systems and Ecology ENSEC'93, Cracow, Poland, July 5–9, pp. 339–350.
Mynarski, S., 1979. Elementy Teorii Systemów I Cybernetyki (Elements of Systems Theory and Cybernetics). Państwowe Wydawnictwo Naukowe, Warszawa (in Polish).
Rosen, R., 1970. Dynamical Systems Theory in Biology. Wiley, New York.
Ruth, M., 1993. Integrating Economics, Ecology and Thermodynamics. Kluwer, Dordrecht.
Schneider, 1987. Schrödinger shortchanged. Nature 328, 300.
Schneider, E., Kay, J., 1993. Energy degradation, thermodynamics and the development of ecosystems. In: Szargut, J., Kolenda, Z., Tsatsaronis, G., Ziębik, A. (Eds.), Proceedings of the International Conference ENSEC'93, Energy Systems and Ecology, vol., Cracov, Poland, July 5–9, pp. 33–42.
Schrödinger, E., 1967. What Is Life? Cambridge University Press, Cambridge, UK.

CHAPTER 2

Further development of thermodynamic views in ecology

2.1 Introduction

Since the year 2000, authors of more and more papers and books begin to apply performance criteria, abandoning thermodynamic optimization, i.e. they develop approaches working with the criteria containing an increasing number of ecological indicators, see: Bakshi and Grubb, 2013; Baumgärtner, 2004; Chen et al., 2002, 2004, 2005, 2010; Chen and Sun, 2004; Hau and Bakshi, 2004; Jørgensen, 2001; Kåberger and Månsson, 2001; Sousa, 2007; Szargut, 2001, 2005; Szargut and Stanek, 2007; Zhu et al., 2003, 2004; and others. The applications stretch from the realm of thermal technology (Szargut, 2001) to chemical and other technologies. Particularly fruitful are ecological applications of exergy for the analysis of cumulative exergy consumption and cumulative exergy losses, the determination of a proecological tax to replace the personal taxes (Szargut, 2002), and the minimization of depletion of nonrenewable resources (Szargut, 2005). Ecological approaches are found in books and papers (Jørgensen, 2001; Upadhyay et al., 2000). In particular, calculations by Upadhyay et al. (2000) indicate that structural complexity is not necessary for dynamical one to exist. Simple ecosystems can display dynamical behavior unpredictable in certain situations. In some cases, when riddled basins are found, even qualitative prediction is denied. Even natural convection, Fig. 2.1, can complicate life.

2.2 Thermodynamics and ecology

Thermodynamics is used increasingly in ecology to understand the integral properties of ecosystems because it is a basic science that describes energy transformation from a holistic viewpoint. In the past three decades, many thermodynamically oriented contributions to the ecosystem theory have appeared; therefore, an important current step toward integrating these contributions is to present them synthetically (Jørgensen, 2001). An ecosystem consists of interdependent living organisms that are also interdependent with their environment, all of which are involved in a continual transfer of energy and mass within a general state of equilibrium or disequilibrium. Thermodynamics can quantify in an exact way how "organized" or "disorganized" a system is, which is useful information to know when trying to understand how a dynamic ecosystem is behaving. As a part of the environmental and ecological modeling

Complexity and Complex Ecological Systems. https://doi.org/10.1016/B978-0-443-19237-1.00007-7

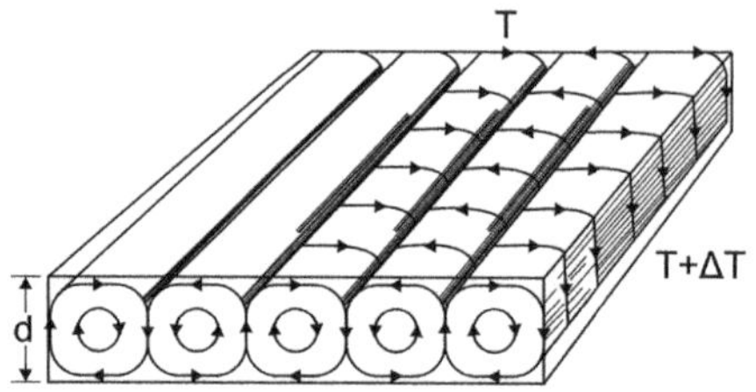

Fig. 2.1
Schematic picture of Rayleigh-Benard convection with streamlines in an ideal roll state. *Based on Cross, W.C., Hohenberg, P.C., 1993. Pattern formation outside of equilibrium. Rev. Mod. Phys. 65(3), 851–1125.*

series, Jørgensen's (2001) book on thermodynamics and ecological modeling is a volume of the current thinking on how an ecosystem can be explained or predicted in terms of its thermodynamic behavior. The core part of the book explains how the thermodynamic theory can specifically be applied to the "measurement" of an ecosystem, including the evaluation of its state of entropy and enthalpy. Additionally, it shows economists how to put the theory to use when trying to quantify the movement of goods and services through a type of complex living system—a human society. A chapter by Bakshi and Grubb (2013) in a book edited by Cabezas and Diwekar (2013) demonstrates the role that thermodynamics can play in assessing the sustainability of technological activities and in improving their design. Since thermodynamics governs the behavior of all macrosystems, it plays a crucial role in understanding physical limits of technologies and for quantifying the contribution of resources. The concept of exergy captures the first and second laws. Exergy is a common currency that flows and is transformed in industrial and ecological systems. This allows for a joint analysis of these systems. This insight permits accounting of the role of ecosystem goods and services in supporting human activities. As ecosystems are critical to sustainability, accounting for their role must be a part of all methods aimed toward the analysis and design of sustainable systems. Thermodynamics provides a rigorous approach for meeting this challenge. In addition, exergy analysis of practical processes and life cycles helps in identifying areas of largest resource inefficiency and opportunities for improvement. This approach complements the insight obtained from assessing the impact of emissions. Case studies based on the life cycle of biofuels and nanomanufacturing are used to demonstrate the important role that thermodynamics can play in sustainability engineering. The reader may also note that a suitable and correct description of the above themes can be achieved in terms of information theory, in view of the link between nonequilibrium thermodynamics and Fisher information (Frieden 1998).

2.3 Thermodynamics and living world

Since Schrödinger's important paper (Schrödinger, 1967) we know that, at first glance, living systems seem to defy the second law of thermodynamics as it insists that, in closed systems, entropy should attain the maximum and disorder should reign. Living systems, however, are the antithesis of such disorder. They display excellent levels of order created from disorder. For instance, plants are highly ordered structures synthesized from disordered atoms and molecules found in atmospheric gases and soils. By turning to nonequilibrium thermodynamics, Schrödinger first recognized that living systems exist well in a world of energy and material and flows (fluxes).

An organism stays alive in its highly organized state by taking energy from outside, from a larger encompassing system, and processing it to produce, within itself, a lower entropy, a more organized state. It is recognized that life constitutes a far-from-equilibrium system that maintains its local level of organization at the expense of the larger global entropy budget. To study living systems from a nonequilibrium perspective requires a reconciliation of biological self-organization and thermodynamics. Furthermore, such a study is expected to yield new principles of physics. We also know that for many proper formulations, the selection and use of the so-called natural variables in the sense of Callen is required, Sieniutycz (2016) TAES.

A theoretical justification for maximum entropy production (MEP) has recently been derived from the information theory, which applies Jaynes' information theory formalism of statistical mechanics to nonequilibrium systems in steady state. He shows that, out of all possible macroscopic stationary states compatible with the imposed constraints (e.g., external forcing, local conservation of mass and energy, and global steady-state mass and energy balance), the state of MEP is selected because it is statistically the most probable, i.e., it is characteristic of the overwhelming majority of microscopic paths allowed by the constraints.

Systems of autocatalytic reactions are a form of positive feedback, where the activity of the system or reaction augments itself in the form of self-reinforcing reactions. An example is a reaction where compound A catalyzes the formation of compound B and B accelerates the formation of A; then, the overall set of reactions is an autocatalytic or positive feedback cycle. Zalewski (2005) enhanced the attractiveness of some of the complexity results by his original study of chaos and oscillations in evolving chemical catalytic systems. Considering two mathematical models, Zalewski and Szwast (2007) found chaos and oscillations in selected biological arrangements of the prey-predator type. Hordijk and Steel (2018) determined the structure of an autocatalytic network as the basis of life's origin and organization, Fig. 2.2. Hau and Bakshi (2004) expanded the classical exergy analysis to apply it for quantitative evaluation of ecosystem products and services.

2.4 The origin of quantification: A brief introduction

Broad exergy applications stretch from the realm of thermal technology (Szargut, 2001) to chemical, metallurgical, and others. Particularly fruitful are ecological applications of exergy for the analysis of cumulative exergy consumption and cumulative exergy losses, the determination of a proecological tax to replace the personal taxes (Szargut, 2002), and the

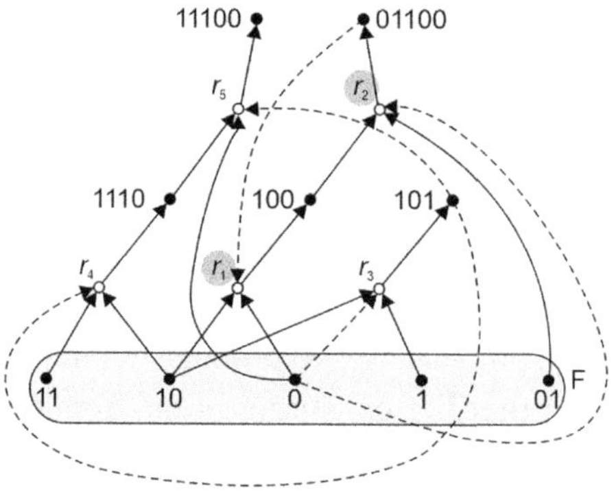

Fig. 2.2

An example of autocatalytc (RAF) set in a polymer model with molecules as "big string polymers" that can be ligated together into longer ones. *Based on Hordijk, W., Steel, M., 2018. Autocatalytic network at the basis of life's origin and organization. Life 8(4), 62.*

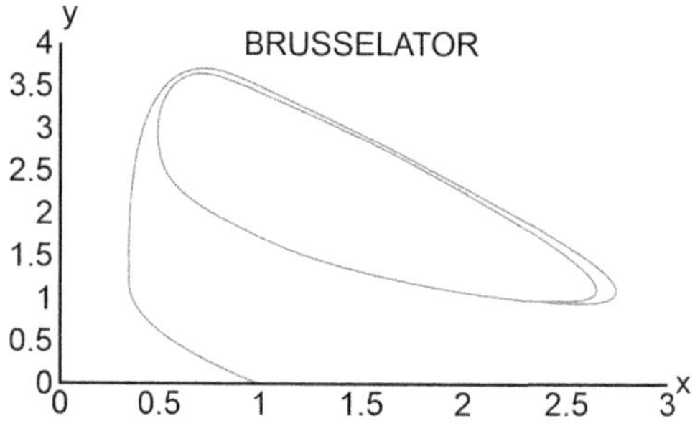

Fig. 2.3

An example of Brusselator may incorporate various solutions of diffusion-free system with enlargements in the vicinity of the fixed point. *Based on Hordijk, W., Steel, M., 2018. Autocatalytic network at the basis of life's origin and organization. Life 8(4), 62.*

minimization of nonrenewable resources depletion (Szargut, 2005). Szargut and Stanek (2007) have developed a thermoecological optimization of a solar collector (Fig. 2.3).

For considerations regarding the appropriateness of FTT optimization criteria and ecological performance functions, see: Bakshi and Grubb, 2013; Baumgärtner, 2004; Chen et al., 2002, 2004, 2005, 2010; Chen and Sun, 2004; Hau and Bakshi, 2004; Jørgensen, 2000, 2001; Kåberger and Månsson, 2001; Sieniutycz and Farkas, 2005; Sieniutycz 2016 (TAES); Sousa, 2007; Szargut, 2005; Szargut and Stanek, 2007; Zhu et al., 2003, 2004; and others.

References

Bakshi, B.R., Grubb, G.F., 2013. Implications of thermodynamics for sustainability. In: Cabezas, H., Diwekar, U. (Eds.), Sustainability: Multi-Disciplinary Perspectives. Bentham Science Publishers, Sharjah, United Arab Emirates, pp. 222–242.

Baumgärtner, S., 2004. Thermodynamic models. In: Proops, J., Safonov, P. (Eds.), Modelling in Ecological Economics. Edward Elgar, Cheltenham, pp. 102–129.

Cabezas, H., Diwekar, U. (Eds.), 2013. Sustainability: Multi-Disciplinary Perspectives. Bentham Science Publishers, Sharjah, United Arab Emirates, pp. 222–242.

Chen, J., Chen, X., Wu, C., 2002. Ecological optimization of a multistage irreversible combined refrigeration system. Energy Convers. Manage. 43, 2379–2393.

Chen, L., Sun, F., 2004. In: Chen, L., Sun, F. (Eds.), Advances in Finite Time Thermodynamics: Analysis and Optimization. Nova Science Publishers, New York.

Chen, L., Zhou, J., Sun, F., Wu, C., 2004. Ecological optimization for generalized irreversible Carnot engines. Appl. Energ. 77 (3), 327–338.

Chen, L., Zhou, J., Sun, F., Wu, C., 2005. Ecological optimization for generalized irreversible Carnot refrigerators. J. Phys. D 38 (1), 113–118.

Chen, L., Xia, D., Sun, F., 2010. Ecological optimization of generalized irreversible chemical engines. Int. J. Chem. Reactor Eng. 8, A121.

Frieden, B.R., 1998. Physics from Fisher Information. Cambridge University Press, Cambridge.

Hau, J.L., Bakshi, B.R., 2004. Expanding exergy analysis to account for ecosystem products and services. Environ. Sci. Technol. 38 (13), 3768–3777.

Hordijk, W., Steel, M., 2018. Autocatalytic network at the basis of life's origin and organization. Life 8 (4), 62.

Jørgensen, S.E., 2000. Application of exergy and specific exergy as ecological indicators of coastal areas. Aquat. Ecosyst. Health Manag. 3 (3), 419–430.

Jørgensen, S.E. (Ed.), 2001. Thermodynamics and Ecological Modelling. Lewis Publishers of CRC Press, Boca Raton.

Kåberger, T., Månsson, B., 2001. Entropy and economic processes—physics perspectives. Ecol. Econ. 36, 165–179.

Schrödinger, E., 1967. What Is Life? Cambridge University Press, Cambridge, UK (first ed. in 1944).

Sieniutycz, S., 2016. Thermodynamic Approaches in Engineering Systems (TAES). Elsevier, Amsterdam, Oxford/ Cambridge, MA.

Sieniutycz, S., Farkas, H. (Eds.), 2005. Variational and Extremum Principles in Macroscopic Systems. Elsevier Science, Oxford.

Sousa, T., 2007. Thermodynamics as a Substantive and Formal Theory for the Analysis of Economic and Biological Systems. PhD thesis carried out at the Department of Theoretical Life Sciences, Vrije Universiteit Amsterdam, The Netherlands and at the Environment and Energy Section, Instituto Superior T´ecnico, Lisbon, Portugal.

Szargut, J., 2001. In: Bilicki, Z., Mikielewicz, J., Sieniutycz, S. (Eds.), Exergy analysis in thermal technology, Ekspertyza KTiSp PAN: Współczesne Kierunki w Termodynamice. Wydawnictwa Politechniki Gdańskiej.

Szargut, J., 2002. Application of exergy for the determination of the proecological tax replacing the actual personal taxes. Energy 27, 379–389.

Szargut, J., 2005. Exergy Method: Technical and Ecological Applications. WIT Press, Southampton.

Szargut, J., Stanek, W., 2007. Thermoecological optimization of a solar collector. Energy 32 (4), 584–590.

Upadhyay, R.K., Iyengar, S.R.K., Rai, V., 2000. Stability and complexity in ecological systems. Chaos Solit. Fractals 11, 533–542.

Zalewski, M., 2005. Chaos and Oscillactions in Chemical Reactors (PhD thesis in Polish). Warsw University of Technology.

Zalewski, M., Szwast, Z., 2007. Chaos and oscillations in chosen biological arrangements of the prey-predator type. Chem. Process Eng. 28, 929–939.

Zhu, X., Chen, L., Sun, F., Wu, C., 2003. The ecological optimization of a generalized irreversible Carnot engine with a generalized heat transfer law. Int. J. Ambient Energy 24 (4), 189–194.

Zhu, X., Chen, L., Sun, F., Wu, C., 2004. The ecological optimization of a generalized Academic Press is an imprint of Elsevier irreversible Carnot engine in the case of another linear heat transfer law. In: Chen, L., Sun, F. (Eds.), Advances in Finite Time Thermodynamics: Analysis and Optimization. Nova Science Publishers, New York, pp. 29–40.

Further reading

Aoki, I., 2001. Entropy and exergy principles in living systems. In: Jorgensen, S.E. (Ed.), Thermodynamics and Ecological Modelling. Lewis (Publishers of CRC Press), Boca Raton, pp. 165–190.

Ayres, R.U., Ayres, L.W., Warr, B., 2003. Exergy, power and work in the US economy 1900–1998. Energy Int. J. 28, 219–273.

Dewar, R., 2003. Information theory explanation of the fluctuation theorem, maximum entropy production and self-organized criticality in non-equilibrium stationary states. J. Phys. A: Math. Gen. 36, 631–641.

Lewandowski, W.M., 2001. Proecological Sources of Renewable Energy. Wydawnictwa Naukowo Techniczne, Warsaw (in Polish).

Schneider, E.D., Sagan, D., 2005. Into the Cool: Energy Flow, Thermodynamics and Life. University of Chicago Press.

Yantovski, Y., 2007. Review of the book: J. Szargut's book: exergy method: technical and ecological applications, WIT Press, Southampton, Boston, 2005. Int. J. Thermodyn. 10, 93–95.

CHAPTER 3

Ascendant perspective of Ulanowicz

3.1 Introduction

An interesting contribution displaying diverse properties of ecological systems, with a few observations and conclusions related to physics, is due to Ulanowicz (2000). His ecological performance functions, initially linked with thermodynamic optimization criteria, expanded in time into more complex and sophisticated structures, describing ecosystems more adequately and with more versatility. Some introductory information on ecosystems can be found in the brief, modest text of the present author (Sieniutycz, 2016).

One of the number of phenomena essential to ecology is that of ecosystem succession, or the more or less repeatable temporal series of configurations that an ecosystem will assume the given habitat after a major disturbance or upon the appearance of a new area. Initially, succession was described in terms of natural history (e.g. Clements, 1916), but more recently ecologists have attempted to describe succession or system development more formally (Odum, 1969). The goal of quantitative ecology is to describe the succession process in purely numerical terms.

The quantification of succession is unlikely to be proved, for despite prevailing temporal regularities, the process is not as deterministic as it was initially portrayed. Clements' almost mechanical description of ecosystem succession was challenged very soon by Gleason (1917), who saw community assembly to be more stochastic by nature (Simberloff, 1980). "Contingencies or random perturbations," are mainly a part of any ecosystem's history, and a quantitative theory of ecosystem succession cannot assume a priori that such chance events will always average out (Ulanowicz's, 2000).

What follows is basically an abbreviation of Ulanowicz's (2000) description of one particular attempt to quantify the process of ecosystem development. The approach fails under the rubric of system ascendancy, so named after the key index spawned by the theory. Ascendancy is applied to gauge the activity and organization inherent in an ecosystem. The approach is neither purely mechanical nor unconditionally stochastic – extremes that to date have characterized most quantitative endeavors in ecosystem science. Rather, the formulation of ascendancy resembles Popper's (1990) call to develop a "calculus of conditional probabilities."

Complexity and Complex Ecological Systems. https://doi.org/10.1016/B978-0-443-19237-1.00006-5

Popper (1990) regards the processes of life as "almost lawful" in the sense that they are guided by sets of "propensities" or generalizations of Newtonian-like inputs, which are constantly being disrupted by contingent events. Chance does not act on individual component processes in isolation; however, it is assumed in the genetic theory (Fisher, 1930). Ecosystem processes, almost by definition, are coupled to one another, the situation allowing for the effects of chance events to be incorporated into the ongoing history of the system. How a chance event affects a process is dependent in part on conditions elsewhere in the system. Whence the need appears to describe chance not in terms of the ordinary statistics common to most of contemporary biology and physics, but in terms of Bayesian, or conditional probabilities (Ulanowicz, 2000).

The game in constructing a broad, quantitative theory of ecosystem development is to focus first upon the agency behind the "law-like" progression toward a developed configuration, and thereafter to quantify the actions of this agency, not in a conventional, deterministic fashion, but in contingent, probabilistic terms that can incorporate historical and nonlocal events (Ulanowicz, 2000). As he states, the majority of researchers commonly claim that life processes are so difficult to explain because they involve highly reflective and ultimately self-entailing behaviors (Rosen, 1991). While negative feedback is the crux of most internal system regulation, theorists acknowledge that the pressures behind the proliferation and evolution of living forms have more to do with positive feedback and with autocatalytic activities in particular (e.g., Eigen, 1971; Haken, 1988; Kauffman, 1995). Before going further, it is necessary to specify more precisely how the term "autocatalysis" is to be used in ecology, Ulanowicz (2000).

3.2 A cause driving the development

Autocatalysis is a special case of positive feedback (DeAngelis et al., 1986). This sort of feedback can arise in various circumstances/scenarios, some of which involve negative interactions. By autocatalysis ecologists mean the positive feedback comprised wholly of positive component interaction. Ulanowicz (2000) presents in Fig. 3.1 a schematic of autocatalysis among three processes. Some scientists require only that the propensities for positive influence be stronger than cumulative decremental interferences. In the face of environmental contingencies, autocatalytic activities behave in ways that transcend mechanism

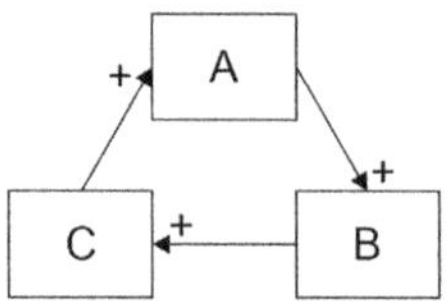

Fig. 3.1
Schematic of hypothetical three-component autocatalytic cycle (Ulanowicz, 2000).

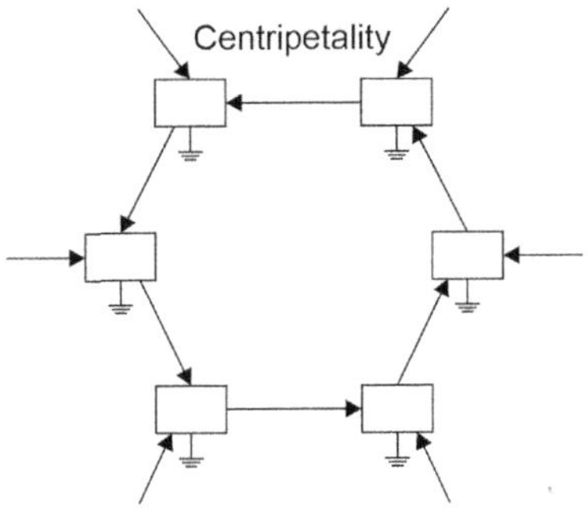

Fig. 3.2
Autocatalytic cycle exhibiting *centripetality* (Ulanowicz, 2000).

(Ulanowicz, 1997), e.g., there is a selection pressure that the overall autocatalytic form exerts upon its components, Ulanowicz (2000).

Unlike Newtonian forces, which always act in equal forward and opposite directions, the selectivity pressure associated with catalysis is inherently asymmetric. Autocatalytic configurations impart a definite sense/direction to the behaviors of systems in which they appear. "They tend to ratchet all participants toward an even greater level of performance" Ulanowicz (2000). The degradation of energy is a spontaneous process governed by the second law of thermodynamics. But it would be erroneous to assume that the autocatalytic loop is itself passive and merely driven by the energy gradient (Ulanowicz, 2000). Any change decreasing the energy intact by a participant would ratchet down activity throughout the loop. Or move by degrees in one direction only, continuing "a ratcheting lopping tool."

Fig. 3.2 of Ulanowicz (2000) shows an autocatalytic cycle exhibiting *centripetality*. As this researcher states: "The asymptotic assemblage behaves as a focus upon which converge increasing amounts of energy and material that the system draws unto itself, cf. Jorgensen (1992)" and in a different text: Contemporary systems ecology has long been occupied with mechanical explanations of behavior; however, the physical theory that undergirds such explanations has certain limits. It's not that the physical force laws are ever violated, but with heterogeneous, irreversible relationships they are subject to subtle/"aleatoric" influences. Therefore, as Ulanowicz (2000) seems to claim (SS): "physical laws can only constrain, but not determine, outcomes. Such complex systems are more adequately treated in the framework of quantified networks of interrelations." The application of simple information theory to networks reveals that "ecosystems cannot achieve maximal efficiency without growing vulnerable to novel disturbances," (Ulanowicz, 2021). The appearance of centripetality and the persistence of form beyond constituents are decidedly non-Newtonian behaviors. Although a living system requires material and mechanical elements, it is evident that some behaviors, especially those on a longer time scale are, to a degree, autonomous of lower-level events (Allen and Starr, 1982). Attempts to predict the course of an autocatalytic configuration by the

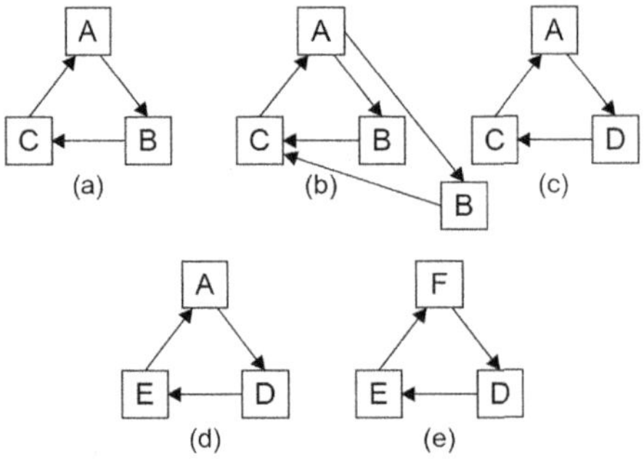

Fig. 3.3
Successive replacement of components in an autocatalytic loop (Ulanowicz, 2000).

ontological reduction to material constituents and mechanical operations are, accordingly, doomed over the long run to failure (Ulanowicz, 2000) (Fig. 3.3).

To recapitulate the contents of sections 2 and 3 in Ulanowicz (2000), we can say: generalized autocatalytic systems can show at least eight behaviors, which, when taken together, mitigate against viewing them as mechanical systems that will make them susceptible to the reductionist analysis. This happens because generalized autocatalysis induces both growth and selection. The autocatalysis exhibits also an asymmetry that can give rise to the "centripetal amassing of material and available energy." The presence of more than a single autocatalytic pathway in a system represents the potential for competition. Such autocatalytic behavior is autonomous, to a degree, of its microscopic constitution. These behavior attributes emerge whenever the scale of observation becomes large enough. The overall effects of generalized autocatalytic behavior are exhibited both extensively as a function of system size and intensively, i.e., independent of size. The former is expressed as an increase in total activity, while the latter resembles the "topological pruning" of processes that participate less efficiently in the autocatalytic activities. The present task is to quantify both aspects of growth and development (Figs. 3.4 and 3.5).

3.3 Quantification of growth and development

To quantify growth, one denotes the magnitude of any transfer of material from any donor (prey) i to its receptor (predator) j by T_{ij}. Then, the one measure of the system activity is the sum of all such exchanges; a quantity referred to in economic theory as the "total system throughput," T, i.e.,

$$T = \Sigma_{ij} T_{ij} \tag{3.1}$$

If calculating the system's "size" by its level of activity, one should recall the common practice in the economic theory, where the size of a country's economy is gauged by its "gross domestic

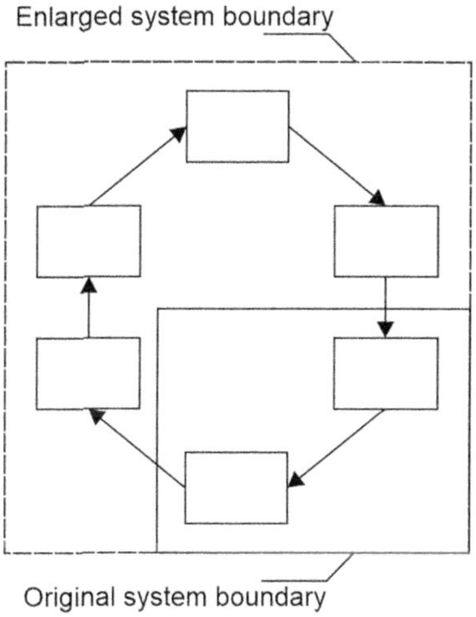

Fig. 3.4
The emergence of nonmechanical behavior as the scope of observation is enlarged (Ulanowicz, 2000).

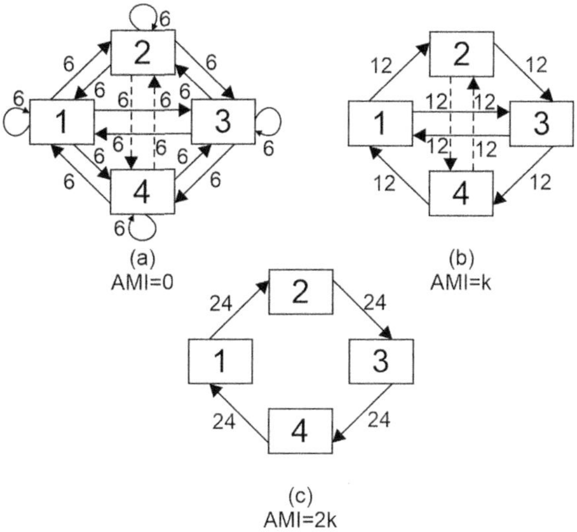

Fig. 3.5
The increase in mutual information as flows become progressively constrained (Ulanowicz, 2000).

product." Quantifying the intensive development is somewhat more difficult. The goal here is to quantify the transition from a very loosely coupled, highly indeterminate collection of exchanges to one in which exchanges are more constrained by autocatalysis to flow along the most efficient pathways. On the way, and with conditional probabilities, an expression should be found for the averaged mutual constraint, A (Ulanowicz, 2000). In terms of the measurable exchanges, the estimated (averaged, mutual) constraint takes the form

$$A = k\Sigma_{ij}\left(T_{jj}/T\right)\log\left[T_{jj}/T\left(\Sigma_k T_{ik}\right)^{-1}\left(\Sigma_l T_{lj}\right)^{-1}\right] \quad (3.2)$$

The above form is equivalent to that of Ulanowicz (2000), presented in Eq. (8).

3.4 System ascendancy

The corresponding system ascendancy expressed in terms of trophic exchanges is

$$A = \Sigma_{ij}T_{jj}\log\left[T_{jj}/T\left(\Sigma_k T_{ik}\right)^{-1}\left(\Sigma_l T_{lj}\right)^{-1}\right] \quad (3.3)$$

To signify that the measure (3.2) has changed its qualitative nature into (3.3), their propagator (Ulanowicz, 1980, 2000) has been chosen to name A of Eq. (3.3) as the system "ascendancy." This quantity measures both the size and the organizational status of the network of exchanges that occur in an ecosystem.

References

Allen, T.F.H., Starr, T.R., 1982. Hierarchy. University of Chicago Press, Chicago.

Clements, F.E., 1916. Plant Succession. An Analysis of the Development of Vegetation. Carnegie Institution of Washington, Washington, D.C.

DeAngelis, D.L., Post, W.M., Travis, C.C., 1986. Positive Feedback in Natural Systems. Springer-Verlag, NY.

Eigen, M., 1971. Selforganization of matter and the evolution of biological macromolecules. Naturwiss 58, 465–523.

Fisher, R.A., 1930. The Genetical Theory of Natural Selection. Oxford University Press, Oxford, UK. 272 pp.

Gleason, H.A., 1917. The structure and development of the plant association. Bull. Torrey Bot. Club. 44, 463–481.

Haken, H., 1988. Information and Self-organization. Springer-Verlag, Berlin. 169 pp.

Jorgensen, S.E., 1992. Integration of Ecosystem Ilen Theories; A Pattern. Kluwer, Dordrecht.

Kauffman, S.A., 1995. At Home in the Universe. The Search for the Laws of Self-Organization and Complexity. Oxford University Press, Oxford.

Odum, E.P., 1969. The strategy of ecosystem development. Science 164, 262–270.

Popper, K.R., 1990. A World of Propensities. Thoenames, Bristol.

Rosen, R., 1991. Life Itself: A Comparative Inquiry into the Nature, Origin, and Foundation of Life. Columbia University Press, NY.

Sieniutycz, S., 2016. Ecology and ecological optimization. In: Thermodynamic Approaches in Engineering Systems. Elsevier, Oxford, pp. 531–536. Sec. 11.5.

Simberloff, 1980. A succession of paradigm in ecology. Essentialism to materialism and prohibilism. Synthese 43, 3–39.

Ulanowicz, R.E., 1980. An hypothesis on the development of natural communities. J. Theor. Biol. 85, 223–245.

Ulanowicz, R.E., 1997. Ecology, the Ascendent Perspective. Columbia University Press, New York.

Ulanowicz, R.E., 2000. Ascendancy: a measure of ecosystem performance. In: Jorgensen, S.E., Muller, F. (Eds.), Handbook of Ecosystem Theories and Management. Lewis Publishers, Boca Raton, USA, pp. 303–315.

Ulanowicz, R.E., 2021. Socio-Ecological Networks: A Lens That Focuses Beyond Physics Frontiers in Ecology and Evolution., p. 9, https://doi.org/10.3389/fevo.2021.643122.

Further reading

Costanza, R., 1992. Toward an Operational Definition of Ecosystem Health. pp. 239–256.

Costanza, R., Norton, B.G., Hankel, B.D. (Eds.), 1992. Ecosystem Health: New Goals. Island Press, Washington, DC. 289 pp.

Feistel, R., Ebeling, W., 1989. Evolution in Complex Systems. Springer. 242 pp.

Field, J.G., Maloney, C.L., Attwood, C.G., 1989. Network analysis of simulated succession after an upwelling event. In: Wulff, F.W., Field, J.G., Mann, K.H. (Eds.), Network Analysis in Marine Ecology: Methods and Applications. Springer-Verlag, Berlin, pp. 132–158.

Holling, C.S., 1986. The resilience of terrestrial ecosystems, local surprise and global change. In: Clark, W.C., Munn, R.E. (Eds.), Sustainable Development of the Biosphere. Cambridge University Press, Cambridge, UK, pp. 297–317.

Magean, M.T., Costanza, R., Ulanowicz, R.E., 1995. The development, testing and application of a quantitative assessment of ecosystem health. Ecosyst. Health 1 (4), 201–213.

Pahl-Wostl, C., 1992. Information theoretical analysis of functional temporal and spatial organization in flow networks. Math. Comput. Model. 16 (3), 35–52.

Tribus, M., McIrvine, E.C., 1971. Energy and information. Sci. Am. 225, 179–188.

Weber, B.H., Depev, D.J., Dyke, C., Salthe, S.N., Schneider, E.D., Ulanowicz, R.E., Wicken, J.S., 1989. Evolution in thermodynamic perspective: an ecological approach. Biol. Philos. 4, 373–405.

Westra, I., 1994. An Environmental Proposal for Ethics: The Principle of Integrity. Rowman & Littlefeld, Lanham, MD.

Jorgensen, S.E., Muller, F. (Eds.), 2000. Handbook of Ecosystem Theories and Management. Lewis Publishers, Boca Raton, USA.

Ulanowicz, R.E., 1986. A phenomenological perspective of ecological development. In: Postom, T.M., Pardy, R. (Eds.), Aquatic Technology and Environmental Fate Ninth Volume, ASTM STP 921. American Society for Testing and Materials, Philadelphia, pp. 73–81.

CHAPTER 4

Genetic diversity and the spread of populations

4.1. Introduction

This chapter follows in its large part the recent results of researchers affiliated with The Ghent University (Mortier et al., 2020) who have shown by extending original diversity ideas of Pielou (1966) that environmental change can move the physiological limits of a range and therefore lead to range expansions as determined by population growth and spread (Chuang and Peterson, 2016). Ranges can alternatively expand beyond the existing geographical limits by the introduction of individuals away from their original range. But alongside the environmental opportunities for range expansions, population spread requires the individual capabilities to do so. Individual-level life history traits related to reproduction and dispersal will influence the extent and variation in population spread and therefore range border dynamics (Fisher, 1937; Angert et al., 2011). As these traits have a genetic basis in many organisms (Roff, 2001; Saastamoinen et al., 2018), range dynamics should to an important extent be determined by the population's genetic composition. Genetic diversity, in numbers and in variation in identity of genotypes, has a well-studied positive effect on various ecological processes. Genetic diversity tends to improve ecological performance as expressed by fitness-associated proxies as higher population growth rates, productivity (e.g., Reusch et al., 2005), or movement (e.g., Wagner et al., 2017). This positive relationship between genetic diversity and a variety of demographic processes can be explained by several mechanisms (Hughes et al., 2008; Bolnick et al., 2011).

Pielou's (1966) basic paper states that the information content may be used as a measure of the diversity of a multispecies biological collection. The diversity of small collections, all of whose members can be identified and counted, is defined by Brillouin's measure of information¸ Brillouin (1956). With larger collections, it becomes necessary to estimate diversity; what is estimated is Shannon's measure of information, which is a function of the population proportions of the several species (Shannon and Weaver, 1969). Different methods of estimation are appropriate for different types of collections. If the collection can be randomly sampled and the total number of species is known, Basharin's formula may be used (Pielou, 1966). With a random sample from a population containing an unknown number of species, Good's method is sometimes applicable (Pielou, 1966). With a patchy population of sessile organisms, such as a plant community, random samples are unobtainable since the contents of a

Complexity and Complex Ecological Systems. https://doi.org/10.1016/B978-0-443-19237-1.00001-6

randomly placed quadrat are not a random sample of the parent population. To estimate the diversity of such a community, a method is proposed whereby the sample size is progressively increased by addition of new quadrats; as this is done the diversity of the pooled sample increases and then levels off. The mean increment in total diversity that results from enlarging the sample still more then provides an estimate of the diversity per individual in the whole population.

A supplementary comment of Ricotta and Avena (2003) points out the relationship between Pielou's evenness and landscape dominance within the context of Hill's diversity profiles. This comment states that entropy-related biodiversity indices deriving their conceptual basis from Shannon's information theory have a long history of use in ecology for quantifying community structure and diversity. In addition, in the last two decades, numerous information-theoretical indices, such as the landscape dominance index, have been extensively applied to characterize landscape diversity in space and time. In their own contribution, Ricotta and Avena (2003) find a simple analytical relation between Pielou's evenness J and landscape dominance D within the broader context of Hill's parametric diversity family. Within this context, Ricotta and Avena (2003) recommend the use of Hill's diversity number evenness $E_{1,0}$ to overcome the shortcomings both of Pielou's evenness J and the landscape dominance index D.

Note, however, that there are many other types of extensions of Pielou (1966) equations, as e.g., those involving periodically forced Pielou equation. The periodic case was investigated by Camouzis and Ladas (2007) who have searched for the global character of solutions of the periodically forced Pielou's equation and have proven a theorem extending to the periodic case the result of the autonomous case. Their result has turned out to be successful provided that a sequence $\{\beta n\}$ exists such that every positive solution of the equation converges to a periodic solution with prime period k with positive values, and some associated inequalities are satisfied.

Population spread from a limited pool of founding propagules is at the basis of biological invasions, Mortier et al. (2020). The size and genetic variation of these propagules eventually affect whether the invasion is successful or not. The inevitable bottleneck at introduction decreases genetic diversity and therefore should affect population growth and spread. However, many heavily bottlenecked invasive populations have been successful in nature. Negative effects of a genetic bottleneck are typically considered to be relaxed in benign environments because of a release from stress. Despite its relevance to understand and predict invasions, empirical evidence on the role of genetic diversity in relation to habitat quality is largely lacking. Mortier et al. (2020) use the mite *Tetranychus urticae* Koch as a model to experimentally assess spread rate and size of genetically depleted inbred populations versus enriched mixed populations. This was assessed in replicated linear patch systems consisting of benign (bean), challenging (tomato), or a gradient (bean to tomato) habitat. As expected, these researchers found no effect of genetic diversity on the population size in benign habitat but found it increased population size in challenging habitat. However, they found that population spread rates were increased due to genetic diversity in the benign but not in the challenging habitat. Additionally, variance in spread was consistently higher in genetically poor populations and highest in the challenging habitat. Their experiment challenges the general view that a bottleneck in genetic variation decreases invasion success in challenging but not benign environments.

Invasive microbial species constitute a major threat to biodiversity, agricultural production, and human health. Over the past century, a multitude of invasive species have emerged as threats to forest and agricultural ecosystems worldwide. Stauber et al. (2021) investigate emergence and diversification of a highly invasive chestnut pathogen lineage across southeastern Europe. Microbial invasions are often dominated by one or a small number of genotypes, yet the underlying factors driving invasions are poorly understood. The chestnut blight fungus *Cryphonectria parasitica* first decimated the North American chestnut, and a more recent outbreak threatens European chestnut stands. To unravel the chestnut blight invasion of southeastern Europe, Stauber et al. (2021) sequenced 230 genomes of predominantly European strains. Genotypes outside of the invasion zone showed high levels of diversity with evidence for frequent and ongoing recombination. The invasive lineage emerged from the highly diverse European genotype pool rather than a secondary introduction from Asia or North America. The expansion across southeastern Europe was mostly clonal and is dominated by a single mating type, suggesting a fitness advantage of asexual reproduction. The findings of the researchers show how an intermediary, highly diverse bridgehead population gave rise to an invasive, largely clonally expanding pathogen.

Environmental change can move the physiological limits of a range and therefore lead to range expansions as determined by population growth and spread (Chuang and Peterson, 2016). Ranges can alternatively expand beyond the existing geographical limits by the introduction of individuals away from their original range. But alongside the environmental opportunities for range expansions, population spread requires the individual capabilities to do so.
Individual-level life history traits related to reproduction and dispersal will influence the extent and variation in population spread and therefore range border dynamics (Fisher, 1937; Angert et al., 2011). As these traits have a genetic basis in many organisms (Roff, 2001; Saastamoinen et al., 2018), range dynamics should to an important extent be determined by the population's genetic composition. Genetic diversity, in numbers and in variation in identity of genotypes, has a well-studied positive effect on various ecological processes. Genetic diversity tends to improve ecological performance as expressed by fitness-associated proxies as higher population growth rates, productivity (e.g., Reusch et al., 2005) or movement (e.g., Wagner et al., 2017). This positive relationship between genetic diversity and a variety of demographic processes can be explained by several mechanisms (Hughes et al., 2008; Bolnick et al., 2011) characterized briefly as follows:

(1) A higher genetic diversity increases opportunities for natural selection to act, hence increasing the average fitness in the population which can eventually increase population growth and improve overall ecological performance. Other evolutionary processes, like inbreeding depression, may in contrast decrease ecological performance.
(2) Enhanced sampling in genetically diverse populations increases the probability of the presence of a phenotype with a positive impact on ecological performance.
(3) A higher genetic diversity increases the variance in phenotypes, which can result in an increase of the mean population's ecological an improved ecological performance relative to the average phenotype when convex relationships exist between genetic diversity and the ecological function (Jensen's inequality principle).
(4) Complementarity effects like niche partitioning and facilitation increase ecological performance by diversifying the ways of performing well.
(5) Portfolio effect stabilize fluctuations in the ecological function. Fluctuations from different genotypes that differ in frequency combine to less fluctuating dynamics.

Range dynamics are even more strongly affected by genetic diversity at longer evolutionary time scales: phenotypes may organize themselves along the range resulting in more dispersive phenotypes disproportionally closer to the leading range edge (Phillips et al., 2010; Burton et al., 2010; Phillips and Perkins, 2019). This spatial sorting and subsequent spatial selection is known to accelerate range expansion (Fronhofer and Altermatt, 2015; Szücs et al., 2017; Van Petegem et al., 2018). Genetic drift during spread may, however, slow expansion (Peischl et al., 2015). These evolutionary processes additionally influence the variability in range expansion in both deterministic and stochastic ways (Williams et al., 2019).

Spread during a biological invasion fundamentally differs from spread from an established range. At introduction, the invading population's genetic diversity is constrained by several bottlenecks during its transport from the ancestral population to form the eventual founding population (Pierce et al., 2017; Renault et al., 2018). The number of founders and the potential admixture of different founding lines (Dlugosch and Parker, 2008; Rius and Darling, 2014) eventually determines the severity of this genetic bottleneck. Some invasive populations even show a higher diversity than their ancestral counterparts (Estoup et al., 2016). The number of invaders also imposes demographic effects. A smaller invading population is more vulnerable to Allee effects (Stephens et al., 1999; Taylor and Hastings, 2005) and demographic stochasticity (Fauvergue et al., 2012) and is therefore less likely to be successful. The number of invaders is often found to increase invasion success (Colautti et al., 2006; Simberloff, 2009; Blackburn et al., 2015). And while the different impact of genetic diversity and demography at establishment has been studied (Ahlroth et al., 2003; Szűcs et al., 2014; Vahsen et al., 2018; Sinclair et al., 2019), its importance for the subsequent population spread is not yet resolved (but see Wagner et al., 2017). Such insights are especially needed to solve the genetic paradox of invasions, i.e., the success of invasions despite severe reductions of genetic diversity

(Dlugosch and Parker, 2008; Mullarkey et al., 2013; Estoup et al., 2016; Schrieber and Lachmuth, 2017).

The environment in which the population spreads is another important ecological driver of invasion success. Introductions can occur in an environment that is similar or vastly different from their ancestral one. When the environment of introduction is different and challenging, evolutionary rescue by means of adaptation (Bell and Gonzalez, 2011) offers a possible route to tackle the imposed challenges and stimulate population growth and spread. In a benign environment, such adaptations may not be needed to attain high population sizes (Schrieber and Lachmuth, 2017). Inbreeding depression is also more strongly manifested in a challenging and stressful environment (Fox and Reed, 2011). Decreased genetic diversity may therefore constrain population spread in stressful but not in benign environments. The difference in importance of genetic diversity for establishment between benign and challenging environments has already been experimentally demonstrated (Hufbauer et al., 2013; Szűcs et al., 2014) and is hinted at in some natural invasions (Daehler and Strong, 1997; Hawley et al., 2005). Environments are, however, seldom homogenous in the environmental parameters that determine a species' niche. Rather they gradually change in an autocorrelated way (Legendre, 1993).

Such environmental gradients are anticipated to affect the rate and success of spread in ways that are different from homogeneous benign and challenging environments. Like for evolutionary processes (Bell and Gonzalez, 2011), a gradual increase in stress may favor population spread compared with a sudden change into a challenging environment and enforce an ecological rescue mechanism.

The impact of evolution on range expansion dynamics in different environments is unpredictable (Williams et al., 2019), but significant insights in range expansion can be gained by studying the immediate effects of genetic diversity on the onset of expansions. The genetic diversity of founders can inform which populations are initially better primed to expand their range, to get ahead and ultimately to invade successfully. The researchers, Mortier et al. (2020), specifically expected genetic diversity to increase population spread more in a challenging environment compared with a benign one. They also explored the effect of genetic diversity on variability in population spread. In parallel with Williams et al. (2019), the deterministic and stochastic forces direction of this effect probably depends on the balance of, affecting the predictability of its effect. Because failed invasions cannot be studied in nature, the researchers established replicated experimental populations of *T. urticae* Koch (two-spotted spider mite) in linear patch systems in which population spread rate and demography were followed for approximately three generations. Spreading populations varied in their level of genetic diversity at the start (single-female lines or mixed lines), independent of the number of introduced individuals, and differed in the kind of environment they were introduced to (benign, challenging or a gradient from benign to challenging) (Figs. 4.1 and 4.2).

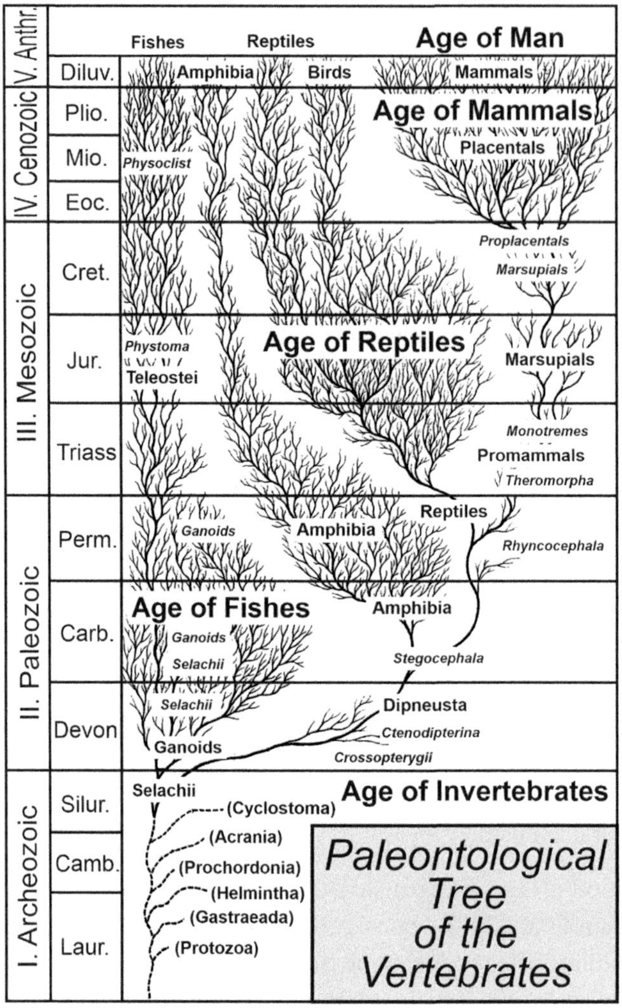

Fig. 4.1

Population spread from a limited pool of founding propagules is at the basis of biological invasions. *Based on Fig. 5.3 in Mortier et al. (2020).*

4.2. Methods

4.2.1 Model system

Mortier et al. (2020) tested population spread of *T. urticae* Koch (two-spotted spider mite), a generalist arthropod herbivore. This mite is a known pest species that has been found on more than 900 hosts all over the world (Navajas, 1998). The species is a model in ecological and evolutionary research because of its ease of use, high fertility, and annotated genome (Belliure et al., 2010; Macke et al., 2011; Van Petegem et al., 2018; Masier and Bonte, 2020). For this

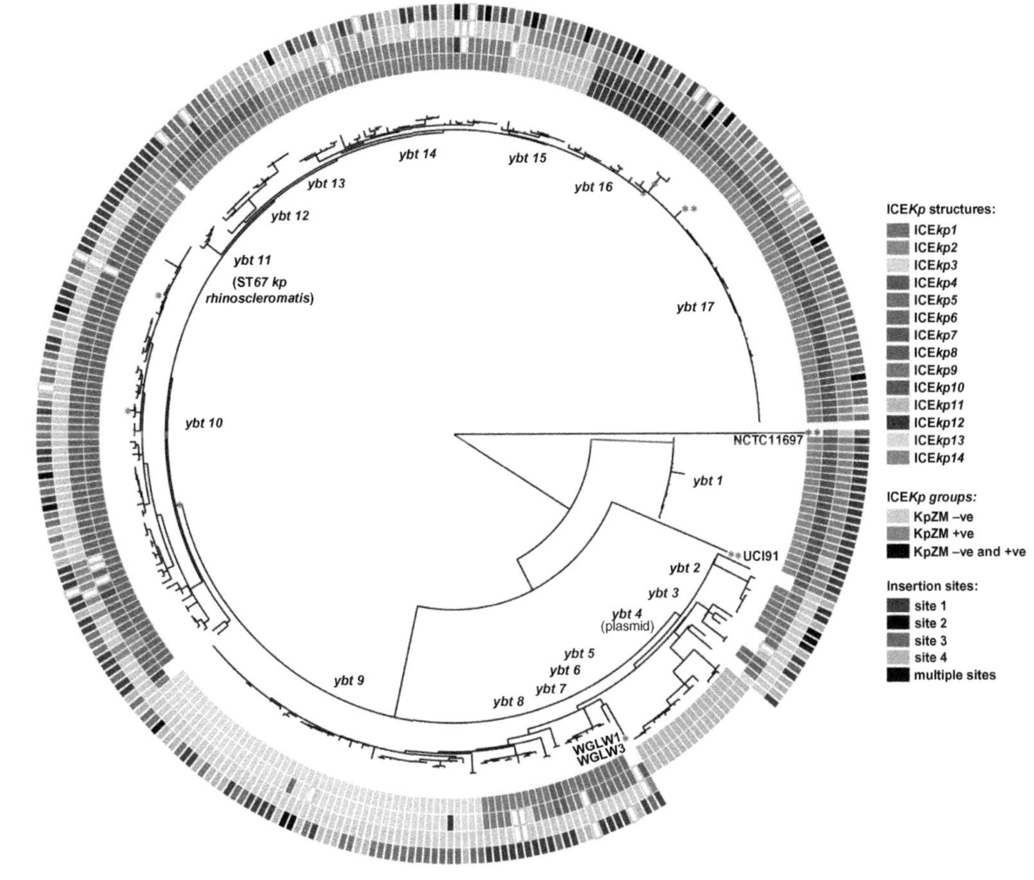

Fig.4.2

A visualization of an external file that holds a picture, illustration. Object name is mgen-4-196-g001.jpg. Recombination-filtered phylogenetic analysis of 329 yersiniabactin sequence types identified across 842 genomes. Comment: Each leaf represents a single yersiniabactin sequence type (YbST), and these YbST sequences cluster into 17 lineages, as labeled. Tracks (from inner to outer): (1) lineage (key as labeled above the tree nodes; white, unassigned), (2) ICEKp structure (white, undetermined), (3) presence or absence of KpZM module and (4) tRNA-Asn insertion site (white, undetermined). Recombination events (Fig. S2) are depicted with a red asterisk next to the relevant branch or a single YbST. *Based on Information of The National Center for Biotechnology (2022).*

experiment, the researchers collected twelve natural populations from a variety of host plants and two laboratory-populations in September 2018 (more information in Appendix S1 of Mortier et al. (2020)). The researchers sampled at least 50 individuals with many more in most of the sampled populations. They maintained the collected populations on *Phaseolus vulgaris* (bean) leaf patches on wet cotton in petri dishes (150 mm diameter) sealed off by a lid with ventilation through which the mites could not escape to keep populations from contaminating each other. They maintained these populations in the laboratory for 6 months, which amounted to around fifteen generations, before the start of the first procedures (creating the mixed lines, see below).

The researchers used *P. vulgaris* var. prelude (bean) plants or leaf patches to keep the stocks on, to perform the single female line procedures and as a benign host in the experiment. Bean is known to be an optimal host that shows little defense against the mites. They used *Solanum lycopersicum* moneymaker (tomato) as a challenging host. The mites occur on tomato but in past experiments, they attained a lower fitness on it (0.2–0.25 of fecundity on bean, Alzate et al., 2017). Local adaptation to tomato is possible but is never observed to result in a higher fitness on tomato compared with bean (Alzate et al., 2017; Mortier and Bonte, 2020). None of the natural populations were sampled from tomato or a host that is taxonomically from the same family (Solanaceae).

4.2.2 Genetic diversity

To obtain starting populations of mites with different levels of genetic variation, Mortier et al. (2020) created one genetically rich population by mixing all collected wild and laboratory lines. Carbonnelle et al. (2007) demonstrated significant genetic differences between natural populations of *T. urticae* in Western Europe at similar geographic scales. The genetically rich mix was formed two months prior to the start of the experiment, spanning around five generations, in order to leave the mixed population enough time to avoid any effect of outbreeding depression. This mixed line was kept in four crates with four to eight bean plants each that were regularly mixed. Each bean plant contained a few hundred to a thousand individuals at all times resulting in a total population of the order of magnitude of ten thousand. This setup supported a high population size to avoid subsequent loss of genetic diversity due to drift or effects of linkage disequilibrium as much as possible. Multiple genetically poor populations were established as single female lines from eight of the collected wild and laboratory lines. The single-female lines were formed by sampling one unfertilized female from a collected line by transferring one quiescent deutonymph, the mite's life stage on the verge of adulthood, to a separate leaf. Due to their parthenogenetic nature, unfertilized eggs from that mite will exclusively produce males that are, then, back-crossed with their mother. As a result, the female produces fertilized eggs hatching males and females. With this procedure, the researchers established a population that consists only of genetic material from the original female, not considering mutations. Low genetic diversity is better retained by smaller population sizes on the leave disks compared to the whole plants. Since densities have been shown to provoke maternal effects on dispersal distance (Bitume et al., 2014), the researchers ensured the populations on whole plants and on leaf discs to be maintained at the 168 same high densities, close to carrying capacity. They performed this procedure with four unfertilized female from each collected line to hedge for likely failure. A mother laid eggs and the unfertilized eggs developed at 26°C, close to the maximum developmental speed. The mother was kept at 17°C while males developed in order to slow the mother's ageing and preserve her fertility to the time of fertilization. During this procedure and in the following four weeks to the start of the experiment, the mites at all times were kept on bean patches (±5cm × 6cm) 174 on wet cotton in a petri dish.

4.2.3 Population spread

Mortier et al. (2020) tested population spread dynamics in experimental linear system containing plant patches that were connected with bridges (their Appendix S2: Fig. S2). Every experimental population was placed in a clean plastic crate (26.5cm × 36.5cm) covered in three layers of wet cotton wool (Rolta soft). Patches of plant leaves (1.5cm × 2.5cm) were connected with one another by a parafilm bridge (1 cm × 8 cm) touching the leaf patch, with the remaining edge aligned with paper towel strips. The wet cotton provided an impenetrable and deadly matrix in between plant patches and provided the cost of moving from one patch to another in the form of mortality risk. The researchers started each population spread test by sampling 40 individuals from a start population and placing them on the first patch in their population spread arena. They added two additional connected patches. Every day they added, if needed, new connected patches to always have two empty patches in front of the furthest occupied patch and every two days they replaced the still unoccupied patches to keep the unoccupied patches at the front fresh and attractive for potential spreading mites. From past experience with similar patch setups, they knew the mites very rarely disperse more than two patches per day under the established densities. Additionally, all plant patches were replaced with a fresh patch once a week. The old patch was placed upside down on the fresh patch for two days in order for mites of all life stages to move from to the fresh patch. Because the old patch always dried quickly, most mites moved to the fresh patch within those two days. Replacing all plant patches replenished resources to sustain the core of the population. These linear patch systems snaked through our crates for twelve possible patches (Appendix S2: Fig. S2 in Mortier et al., 2020).

In the case where a thirteenth (and subsequent) patch was needed, the first (and subsequent) patch and its connection to the next was removed to make space for the new one. The researchers sacrificed trailing patches since they were mostly interested in the leading edge dynamics. The arenas were kept at room temperature, around 23°C, with a 16:8 h L:D photoperiod.

They started population spread tests in three environments: (1) a benign environment of all bean patches, (2) a challenging environment of all tomato patches, and (3) a gradient from benign to challenging patches (with 3 bean patches, 1 tomato patch, 2 bean patches, 2 tomato patches, 1 bean patch, 3 tomato patches, 1 bean patch followed by all tomato patches). They started sixteen population spread tests in each environment. Eight tests of a genetically poor population were each started from a single female line of a different natural population. Eight tests of a genetically rich population were each started from mites from a different plant in the mixed population. The researchers recorded population spread as the furthest occupied patch in each range on a daily basis and recorded the number of mites on each patch on a weekly basis over the duration of 35 days or 5 weeks. Though generations started to overlap, they estimated that this amounted to around three generations.

4.2.4 Statistical models

Mean population spread

Mortier et al. (2020) analyzed the outcome of their experiment using Bayesian inference. They applied advanced Bayesian multilevel modeling with the R Package brms (Bürkner, 2018) and package implements "Stan" (Carpenter et al., 2017) as a framework for parameter posterior estimation using Hamiltonian Monte Carlo (HMC). They constructed a multilevel model to estimate effects on mean population spread. They modeled furthest occupied patch as response variable with a Gaussian distribution. They also modeled a fixed effect of time, the environmental treatment and the genetic diversity treatment and a variable intercept and slope (in time) for each population spread arena (i.e., random effect of replicate population spread arena and its interaction with time).

Variability in population spread

Mortier et al. (2020) estimated effects on variance in population spread among replicates from the same treatment for each point in time. They again modeled population spread (furthest occupied patch) with a Gaussian distribution, but estimated effects on both the mean and standard deviation. They modeled the mean of the Gaussian distribution with a fixed effect of time, the environmental treatment, and the diversity treatment.

However, they modeled no variable intercepts but pooled that variation around the mean originating from among different replicates together with the residual variation in order to model all variation around the mean. They then proceeded to model the standard deviation of the Gaussian distribution also with a fixed effect of time, the environmental treatment in and the diversity treatment in the same model. This way, the model estimated the standard variation for each treatment at each point in time. They calculated coefficients of variance from the estimated mean and standard deviation in population spread.

Total population size

Population size can help explain population spread. Therefore, Mortier et al. (2020) modeled total population size across all patches as a response variable with a negative binomial distribution. They modeled a fixed effect of time, the environmental treatment, and the diversity treatment and a variable intercept and slope (in time) for each population spread arena.

Population density

Mortier et al. (2020) modeled total population sizes across all patches as a response variable with a negative binomial distribution. They modeled a fixed effect of the amount of occupied patches, the environmental treatment, and the diversity treatment. They built a model like the previous but with the amount of occupied patches as fixed effect to understand whether larger populations were larger because they occupied more patches or because they had a higher density.

Mortier et al. (2020) reported model outcomes as plots of the posterior distribution or direct calculations of marginal effects of the posterior distribution. Bayesian inference considers less the most likely parameter or marginal effect value but rather the whole distribution of likely values, which can most faithfully be reported visually instead of as a metric. Mortier et al. (2020) always plotted the 0.09 and 0.91 percentiles of the likelihood distribution. They stress that this is by no means a significance threshold but only serves as two arbitrary extremes to aide interpretation of the plotted distribution. They also include the estimated difference between two groups differing in a treatment to directly assess the estimated effect of that treatment in most plots.

A more detailed description of the statistical models and their outcomes can be found in their supplementary materials (their Appendix S3). The data files and scripts to analyze them were published (https://doi.org/10.5281/zenodo.4025183).

4.3. Results

4.3.1 Mean population spread

All populations showed an initial burst of spread during the first days where the relative high density of mites at the starting patch incentivized mites to leave for the next patch. This resulted in an estimated intercept higher than one, the expected intercept as all populations started with one occupied patch (see researchers' Fig. 1, top). Mites in the tomato environment spread very little after this initial burst resulting in the edge lingering around the fourth patch on average. In the bean and gradient environment, however, the population on average kept spreading for the duration of the experiment. The higher genetic diversity of the mixed populations resulted in a faster spreading population in the bean environment as seen in the predominantly negative estimated difference in slope between both diversity treatments (sfl-mix, their Fig. 1, bottom left). Genetic diversity had no convincing effect in the gradient or tomato environment as seen in the estimated differences in slope (sfl-mix) around zero (Fig. 1, ibid, bottom middle, bottom right). On bean, mixed lines reached on average the ninth patch while single female lines reached on average the sixth patch. On the gradient, both treatments reached on average the eighth patch. Contrary to expectations, the effect of genetic diversity on population spread was larger in the benign rather than in the challenging environment.

4.3.2 Variability in population spread

The genetically rich mixed lines had a less variable population spread compared with the single female lines in all environments (see researchers Fig. 2). In the bean and gradient environments, variance was more or less constant and the difference in variance due to genetic diversity was relatively small. However, in the tomato environment, the coefficient of variance increased in time and shows a relatively sizeable difference in the coefficient of variance between genetically poor and rich lines.

4.3.3 Total population size

Total population sizes were smaller on tomato compared with the bean and gradient treatment (their Fig. 3, top). On tomato, Mortier et al. (2020) found a larger population size of mixed compared with single female lines (their Fig. 3, bottom right). The estimated differences between the genetically diverse and depleted lines on bean had zero (i.e., the no difference), close to the 91 percentile visual aide they plot their Fig. 3, left). They estimated no difference on the gradient (Fig. 3, bottom middle).

4.3.4 Population density

By adding the number of occupied patches as a predictor to the model instead of time, they found total population size to be affected by the amount of patches: total population size increased with plant patches (Mortier et al., 2020, Fig. 4, top). Population sizes for a given number of plant patches of mixed and single female lines were estimated not to differ on bean and the gradient (their Fig. 4, bottom left, middle). However, mixed lines showed a higher population size for a given number of plant patches on tomato, as shown by the increasingly positive difference in population size (their Fig. 4, bottom right). This indicated a higher density in mixed lines than in single female lines. Densities of many mixed lines on tomato even exceeded densities in the benign environment.

4.4. Discussion

The hypothesis of genetic diversity benefiting population size and spread in challenging but not in benign environments is only partially validated by the experiment. In contrast to common predictions, Mortier et al. (2020) documented a positive effect of genetic diversity on population spread rate in the benign (bean) environment but not in the challenging (tomato) environment. In support, they detected a positive effect of genetic diversity on total and local population sizes in the challenging tomato environment, but not in the benign bean environment. This accords with findings of Hufbauer et al. (2013) and Szűcs et al. (2014) for population growth in whiteflies and flour beetles.

Ghent (1991) describes his insights into diversity and niche breadth analyses from exact small-sample tests of the "equal abundance hypothesis." Exact test procedures are detailed for niche breadth as the general case, and for diversity as a subset of niche breadth analysis. Numerators of niche breadth probabilities are found as chain multiplications of three individually simple combinatorial calculations, only two of which are needed in analyses of species diversity. Algebraic checks on counts of terms and sums of numerators are developed. The two-niche case is used to demonstrate the high probabilities associated with high niche diversity, and these effectively limiting tests for significant nonrandomness to the low diversity tail. Examination of diversity indices as rank-ordering devices shows that all the common indices (Shannon's H', Brillouin's H, and their close relatives) yield rank orders identical to some algebraic functions of the equiprobable 1:1:...:1 model no matter what alternative model one may claim as his intended interest. These common indices are obtained to yield a very similar rank of orderings, so that the likelihood of disagreement with a correct or perfect ordering is less than 5% for neighboring partitions of N organisms into n niches or s species, and much less than 5% for randomly selected partitions, even if a diversity index is selected at random without regard for its "theoretical suitabilities." The statistical expectation of Simpson's C' (the complement of Simpson's original index) is obtained to be $(n - 1)/n$ for n niches regardless of N, and to be $(s - 1)/s$ for s species, to four decimal places or more, if species diversity sampling is continued until each species is represented on average by 7.5 "individuals" or more. The "evenness" measures V' and V representing rates of observed to maximum values of diversity indices are found to attain values of 0.99 or more at the 50th percentiles of the probability distributions of diversity indices and are judged to be ecologically misleading and valueless. The claim that there is anything about Brillouin's H that suits it more to a census than a sample is disputed. Simpson's C' is recommended as the most serviceable of the examined indices, both because of the simple formulae for its statistical expectations and because its rank of orderings are found identical to those given by the variance or its square root of the counts of species or niche occupants Ghent (1991). Note that the concept of diversity is employed in his work in the sense in which Pielou (1966) and some other researchers defined it to represent a "blend of diversity measures," and his presentation is exceptional clarity.

In line with Wagner et al. (2017), the researchers found that diversity accelerated spread in the benign environment. Such increased expansions should theoretically result from increased population growth and dispersal (Fisher, 1937; Angert et al., 2011), or any combination of these two. Diversity effects are known to affect population growth (Hughes et al., 2008) and can lead to a fast population growth to or even beyond carrying capacity. At this population size, a positive density-dependence of dispersal (Matthysen, 2005; Bowler and Benton, 2005; Kawasaki et al., 2017) can circumvent strong density regulation by competition as a result from strong growth and lead to a faster range expansions (so-called pushed range expansions; Dahirel et al., 2020). Average local population sizes in their experiment did not differ conclusively between the diversity treatments, whereas those of spread did.

Mortier et al. (2020) therefore attribute the accelerated range expansion in part to the fact that diverse population may impose higher per capita dispersal rates by, for instance, a sampling effect that enables more dispersive genotypes to be present or heterosis and positive body condition effects on dispersal (Wagner et al., 2017).

In contrast to the benign environment, genetically diverse and less diverse populations both stop expanding after the initial leap in spread in the challenging environment. This means that less diverse populations spread as fast as diverse populations under these conditions, while reaching lower population sizes. Negative effects of a low genetic diversity on reproduction are therefore anticipated to be compensated by an increase in dispersal. Dispersal to avoid kin competition is well-documented in *T. urticae* (Bitume et al., 2013; Van Petegem et al., 2018) and is a plausible cause of the enhanced dispersal in genetically impoverished populations. Mixed and single female lines reach equal high densities in the benign environment, and any kin competition is likely overruled by resource competition, which reverses a negative effect of diversity on dispersal to a positive.

Population dynamics on the gradient share characteristics from both the homogeneous benign bean and homogeneous challenging tomato environments. Ranges on the gradient reached on average the position where the mites started to encounter more tomato patches than bean patches, with some populations spreading beyond this point. But because population spread started in the benign environment, populations on the gradient unsurprisingly grew and spread at similar rates to populations on bean. No diversity effects were however observed. The genetically poor populations spread as far as the enriched populations on the gradient, but further than genetically poor populations on bean. Mortier et al. (2020) therefore attribute the spread of genetically poor populations to the environment rather than to a diversity effect. The insertion of challenging patches likely stimulated dispersing mites to skip the challenging patches in search of the next bean patch, which suggest an additional effect of the gradient's patchy nature. Obviously, informed movement leading to habitat choice must enable such a behavior (Egas and Sabelis, 2001, Mortier and Bonte, 2020).

The impact of spatial sorting, selection, and local adaptation on range expansions, has been studied experimentally in recent years (Fronhofer and Altermatt, 2015; Van Petegem et al., 2016, 2018; Szűcs et al., 2017). Mortier et al. (2020) deliberately limited the experiment to roughly 2–3 generations to focus on the immediate effect of genetic diversity on the initial population spread. Selection as a precursor of local adaptation and spatial evolution may play an important role from the first generation onward (Szűcs et al., 2017). All replicate populations spread at least a few patches and assuming some heritable difference in the ability to spread, this should result in some spatial sorting while determining the resulting population spread rate at the same time. Similarly, the ability to better reproduce in the local environment, if heritable, will be proportionally overrepresented in the next generation and may increase population spread rate simultaneously. It is therefore clear that if such mechanisms, or any other long-term mechanism, would have manifested during the short term of the experiment, we are actually underestimating them.

Genetically diverse lines were characterized by a lower variability in spread among replicates in every environment, hence showing a more consistent and predictable spread rate. This consequence of genetic diversity is found for many ecological processes (Hughes et al., 2008). The difference between genetically diverse and poor populations is larger and starts earlier in time on the challenging tomato. Williams et al. (2019) attribute variability in range expansion dynamics to the balance of variance generating (stochastic) and variance reducing (deterministic) evolutionary forces. Analogously, Mortier et al. (2020) propose that different diversity effects can have a variance increasing or decreasing effect on population spread. A probably relevant effect is that higher population sizes reduce demographic stochasticity as a consequence. Since movement is positively related to density, and hence population size, variability in spread is also expected to be minimized in genetically diverse populations. This explains differences in variability between the diversity treatments on tomato and explains the higher overall variability on tomato compared with the other environments. The researchers furthermore speculate that larger differences in variability between genetically diverse and poor replicates on tomato are due to specific genotypes disproportionally impacting range spread. Population spread in the genetically poor lines will be either high or low depending on the presence or absence of the impactful genotype. In comparison, all genetically rich lines are more likely to contain this or any other high impact genotype, hence reducing any variance in population spread among replicates. The smaller difference in variability between genetically diverse and poor lines in the benign bean environment suggests a smaller contribution of specific high-dispersive genotypes, but rather a larger contribution of positive effects of diversity in and of itself. The data obtained by Mortier et al. (2020) do not allow them to infer a specific mechanism but they speculate that niche partitioning of residents and dispersers is at play. They are not talking about niche partitioning in terms of type of resource but in terms of diverging movement strategies to exploit the same limited resources in the landscape (Bonte et al., 2014). It is to be expected that competitive subordinates move as an adaptive behavior when competition is strong, as expected in a benign environment (stabilizing mechanisms; Chesson, 2000). Such fitness stabilizing mechanisms can accelerate range spread in genetically diverse populations, not because of the presence of a specific genotype but because the diversity of genotypes provoking synergies that cannot be reached in isolation. It is not unlikely that a possible higher diversity of movement strategies may have induced some strategies to disperse more to outrun competition in the benign environment.

The views that invaded environments are those where species are released from stress is one of the most important paradigms in invasion biology. The enemy-release hypothesis, for instance, builds on this view of invaded environments being inherently benign (Keane, 2002; Colautti et al., 2004). Such a mechanism implies that the release of stressful interactions should outweigh any potential maladaptation to the new environment. Mortier et al. (2020) do not see any reason why this assumption should hold. Rather, a strong bias exists in invasion biology as all observed invasions have been by definition successful, with any reference to failed invasions in both challenging and benign environments lacking. Experiments like these that observe aspects of ongoing invasion of spread instead of the aftermath circumvents this survivorship bias. Another aspect of the classic view on invasions is that high propagule pressure, which avoids demographic stochasticity and results in a higher genetic diversity, increases an invader's success. The results from challenging environments, however, illustrate that a genetic bottleneck does not decrease population spread in the initial phase and, therefore, may ultimately not decrease invasion success. This is opposite to the established explanation for the genetic paradox of invasions that a genetic bottleneck should still affect a population in a challenging environment. However, the results still support the idea that the effect of a genetic bottleneck is conditional on the environment. Bear in mind that these challenging environments already support a lower invasion success inherently that, as it was shown, may not be aided by genetic diversity. Genetic adaptation to the challenging environment may, however, impose a niche shift on the longer term, and therefore allowing the invading population to capitalize on its genetic diversity. Such niche shifts have been observed in experimental (Szücs et al., 2017) and successful natural invasions (Broennimann et al., 2007), but tend to be the exception rather than the rule (Peterson, 2011; Petitpierre et al., 2012; Strubbe et al., 2013).

Appendix

Information on the graphic material associated with the contents of this chapter (refer to the publication by Mortier, F., Masier, S., Bonte D, 2020 in *Ecology*, https://doi.org/10.1002/ecy.3345).

Mortier, F., Masier, S., Bonte, D., 2020. Figures:

Fig. 1: Top: population spread as the furthest occupied patch over the duration of the experiment for mixed (red) and single female (blue) populations in the bean (left), gradient (middle) or tomato (right) environment. The fine lines show the recorded spread of each population for each day while the wide lines with shades represent the statistical (BMC) model estimate with likelihood interval of the 0.09 and 0.91 quantiles. Bottom: differences in estimated slopes of population spread in time (single female line-mix) in the bean (left), gradient (middle) or tomato (right) environment. The dashed line indicates equal estimated slopes.

Fig. 2: Population spread variability as coefficient of variance (CV) of the number of occupied patches over the duration of the experiment for mixed (red) and single female (blue) populations in the bean (left), gradient (middle), or tomato (right) environment. The dots show the calculated coefficient of variance in population spread (standard deviation in population spread divided by mean population spread) among all populations each day while the wide lines with shades represent the statistical (BMC) model estimate with likelihood interval of the 0.09 and 0.91 quantiles.

Fig. 3: Total population size (top) for mixed (red) and single female (blue) populations and estimated differences in total population size between single female lines and mixes (sfl-mix) (bottom) in the bean (left), gradient (middle), or tomato (right) environment. The fine lines show the recorded total population size of each population each week while the wide lines with shades represent the statistical (BMC) model estimate with likelihood interval of the 0.09 and 0.91 quantiles. The estimated differences in total population size (bottom) between single female lines and mixes (sfl-mix) visualizes whether the differences due to genetic diversity in its respective top panel differs from zero (dashed line).

Fig. 4: Total population size regressed over the furthest occupied patch by that population (top) for mixed (red) and single female (blue) populations and estimated differences in total population size between single female lines and mixes (sfl-mix) (bottom) in the bean (left), gradient (middle), or tomato (right) environment. The fine lines show the recorded total population size of each population each week while the lines and shades represent the statistical (BMC) model estimate with confidence likelihood of the 0.09 and 0.91 quantiles. The estimated differences in total population size (bottom) between single female lines and mixes (sfl-mix) visualizes whether the differences due to genetic diversity in its respective top panel differs from zero (dashed line). All recorded time points are plotted for every population, and the model only estimates the range of occupied patches that were observed per combination of diversity and environment.

References

Ahlroth, P., Alatalo, R.V., Holopainen, A., Kumpulainen, T., Suhonen, J., 2003. Founder population size and number of source populations enhance colonization success in waterstriders. Oecologia 137, 617–620.

Alzate, A., Bisschop, K., Etienne, R.S., Bonte, D., 2017. Interspecific competition counteracts negative effects of dispersal on adaptation of an arthropod herbivore to a new host. J. Evol. Biol. 30, 1966–1977.

Angert, A.L., Crozier, L.G., Rissler, L.J., Gilman, S.E., Tewksbury, J.J., Chunco, A.J., 2011. Do species' traits predict recent shifts at expanding range edges? Ecol. Lett. 14, 677–689.

Bell, G., Gonzalez, A., 2011. Adaptation and evolutionary rescue in metapopulations experiencing environmental deterioration. Science 332, 1327–1330.

Belliure, B., Montserrat, M., Magalhaes, S., 2010. Mites as models for experimental evolution studies. Acarologia 50, 513–529.

Bitume, E.V., Bonte, D., Ronce, O., Bach, F., Flaven, E., Olivieri, I., Nieberding, C.M., 2013. Density and genetic relatedness increase dispersal distance in a subsocial organism. Ecol. Lett. 16, 430–437.

Bitume, E.V., Bonte, D., Ronce, O., Olivieri, I., Nieberding, C.M., 2014. Dispersal distance is influenced by parental and grand-parental density. Proc. R. Soc. B Biol. Sci. 281, 20141061.

Blackburn, T.M., Lockwood, J.L., Cassey, P., 2015. The influence of numbers on invasion success. Mol. Ecol. 24, 1942–1953.

Bolnick, D.I., Amarasekare, P., Araújo, M.S., Bürger, R., Levine, J.M., Novak, M., Rudolf, V.H.W., Schreiber, S.J., Urban, M.C., Vasseur, D.A., 2011. Why intraspecific trait variation matters in community ecology. Trends Ecol. Evol. 26, 183–192.

Bonte, D., De Roissart, A., Wybouw, N., Van Leeuwen, T., 2014. Fitness maximization by dispersal: evidence from an invasion experiment. Ecology 95, 3104–3111.

Bowler, D.E., Benton, T.G., 2005. Causes and consequences of animal dispersal strategies: relating individual behaviour to spatial dynamics. Biol. Rev. 80, 205–225.

Brillouin, L., 1956. Science and Information Theory. Academic Press, New York.

Broennimann, O., Treier, U.A., Müller-Schärer, H., Thuiller, W., Peterson, A.T., Guisan, A., 2007. Evidence of climatic niche shift during biological invasion. Ecol. Lett. 10, 701–709.

Bürkner, P.-C., 2018. Advanced Bayesian multilevel modeling with the R package BRMS. R J. 10, 395.

Burton, O.J., Phillips, B.L., Travis, J.M.J., 2010. Trade-offs and the evolution of life-histories during range expansion. Ecol. Lett. 13, 1210–1220.

Camouzis, E., Ladas, G., 2007. Periodically forced Pielou's equation. J. Math. Anal. Appl. 333, 117–127.

Carbonnelle, S., Hance, T., Migeon, A., Baret, P., Cros-Arteil, S., Navajas, M., 2007. Microsatellite markers reveal spatial genetic structure of *Tetranychus urticae* (Acari: Tetranychidae) populations along a latitudinal gradient in Europe. Exp. Appl. Acarol. 41, 225–241.

Carpenter, B., Gelman, A., Hoffman, M.D., Lee, D., Goodrich, B., Betancourt, M., Brubaker, M., Guo, J., Li, P., Riddell, A., 2017. Stan: a probabilistic programming language. J. Stat. Softw. 76.

Chesson, P., 2000. Mechanisms of maintenance of species diversity. Annu. Rev. Ecol. Syst. 31, 343–366.

Chuang, A., Peterson, C.R., 2016. Expanding population edges: theories, traits, and trade-offs. Glob. Chang. Biol. 22, 494–512.

Colautti, R.I., Grigorovich, I.A., MacIsaac, H.J., 2006. Propagule pressure: a null model for biological invasions. Biol. Invasions 8, 1023–1037.

Colautti, R.I., Ricciardi, A., Grigorovich, I.A., MacIsaac, H.J., 2004. Is invasion success explained by the enemy release hypothesis? Ecol. Lett. 7, 721–733.

Daehler, C.C., Strong, D.R., 1997. Reduced herbivore resistance in introduced smooth cordgrass (*Spartina alterniflora*) after a century of herbivore-free growth. Oecologia 110, 99–108.

Dahirel, M., Bertin, A., Haond, M., Blin, A., Lombaert, E., Calcagno, V., Fellous, S., Mailleret, L., Vercken, E., 2020. Shifts from pulled to pushed range expansions caused by reductions in connectedness. bioRXiv.

Dlugosch, K.M., Parker, I.M., 2008. Founding events in species invasions: genetic variation, adaptive evolution, and the role of multiple introductions. Mol. Ecol. 17, 431–449.

Egas, M., Sabelis, M.W., 2001. Adaptive learning of host preference in a herbivorous arthropod. Ecol. Lett. 4, 190–195.

Estoup, A., Ravigné, V., Hufbauer, R., Vitalis, R., Gautier, M., Facon, B., 2016. Is there a genetic paradox of biological invasion? Annu. Rev. Ecol. Evol. Syst. 47, 51–72.

Fauvergue, X., Vercken, E., Malausa, T., Hufbauer, R.A., 2012. The biology of small, introduced populations, with special reference to biological control. Evol. Appl. 5, 424–443.

Fisher, R.A., 1937. The wave of advance of advantageous genes. Ann. Eugenics 7, 355–369.

Fox, C.W., Reed, D.H., 2011. Inbreeding depression increases with environmental stress: an experimental study and meta-analysis. Evolution 65, 246–258.

Fronhofer, E.A., Altermatt, F., 2015. Eco-evolutionary feedbacks during experimental range expansions. Nat. Commun. 6, 6844.

Ghent, A.W., 1991. Insihts into diversity and niche breadth analyses from exact small-sample tests of the equal abundance hypothesis. Am. Midl. Nat. 126, 213–255.

Hawley, D.M., Hanley, D., Dhondt, A.A., Lovette, I.J., 2005. Molecular evidence for a founder effect in invasive house finch (*Carpodacus mexicanus*) populations experiencing an emergent disease epidemic. Mol. Ecol. 15, 263–275.

Hufbauer, R.A., Rutschmann, A., Serrate, B., Vermeil de Conchard, H., Facon, B., 2013. Role of propagule pressure in colonization success: disentangling the relative importance of demographic, genetic and habitat effects. J. Evol. Biol. 26, 1691–1699.

Hughes, A.R., Inouye, B.D., Johnson, M.T.J., Underwood, N., Vellend, M., 2008. Ecological consequences of genetic diversity. Ecol. Lett. 11, 609–623.

Kawasaki, K., Shigesada, N., Iinuma, M., 2017. Effects of long-range taxis and population pressure on the range expansion of invasive species in heterogeneous environments. Theor. Ecol. 10, 269–286.

Keane, R., 2002. Exotic plant invasions and the enemy release hypothesis. Trends Ecol. Evol. 17, 164–170.

Legendre, P., 1993. Spatial autocorrelation: trouble or new paradigm? Ecology 74, 1659–1673.

Macke, E., Magalhaes, S., Bach, F., Olivieri, I., 2011. Experimental evolution of reduced sex ratio adjustment under local mate competition. Science 334, 1127–1129.

Masier, S., Bonte, D., 2020. Spatial connectedness imposes local- and metapopulation-level selection on life history through feedbacks on demography. Ecol. Lett. 23, 242–253.

Matthysen, E., 2005. Density-dependent dispersal in birds and mammals. Ecography 28, 403–416.

Mortier, F., Bonte, D., 2020. Trapped by habitat choice: ecological trap emerging from adaptation in an evolutionary experiment. Evol. Appl. 13, 1877–1887.

Mortier, F., Masier, S., Bonte, D., 2020. Genetically diverse populations spread faster in benign but not in challenging environments. Ecology. https://doi.org/10.1101/2020.11.27.400820. Now published in *Ecology*. doi: 10.1002/ecy.3345.

Mullarkey, A.A., Byers, D.L., Anderson, R.C., 2013. Inbreeding depression and partitioning of genetic load in the invasive biennial *Alliaria petiolata* (Brassicaceae). Am. J. Bot. 100, 509–518.

Navajas, M., 1998. Host plant associations in the spider mite *Tetranychus urticae* (Acari: Tetranychidae): insights from molecular phylogeography. Exp. Appl. Acarol. 22, 201–214.

Peischl, S., Kirkpatrick, M., Excoffier, L., 2015. Expansion load and the evolutionary dynamics of a species range. Am. Nat. 185, E81–E93.

Van Petegem, K.H.P., Boeye, J., Stoks, R., Bonte, D., 2016. Spatial selection and local adaptation jointly shape life-history evolution during range expansion. Am. Nat. 188, 485–498.

Van Petegem, K., Moerman, F., Dahirel, M., Fronhofer, E.A., Vandegehuchte, M.L., Van Leeuwen, T., Wybouw, N., Stoks, R., Bonte, D., 2018. Kin competition accelerates experimental range expansion in an arthropod herbivore. Ecol. Lett. 21, 225–234.

Peterson, A.T., 2011. Ecological niche conservatism: a time-structured review of evidence. J. Biogeogr. 38 (817–827), 542.

Petitpierre, B., Kueffer, C., Broennimann, O., Randin, C., Daehler, C., Guisan, A., 2012. Climatic niche shifts are rare among terrestrial plant invaders. Science 335, 1344–1348.

Phillips, B.L., Brown, G.P., Shine, R., 2010. Life-history evolution in range-shifting populations. Ecology 91, 1617–1627.

Phillips, B.L., Perkins, T.A., 2019. Spatial sorting as the spatial analogue of natural selection. Theor. Ecol. 12, 155–163.

Pielou, E.C., 1966. The measurement of diversity in different types of biological collections. J. Theor. Biol. 13, 131–144.

Pierce, A.A., Gutierrez, R., Rice, A.M., Pfennig, K.S., 2017. Genetic variation during range expansion: effects of habitat novelty and hybridization. Proc. R. Soc. B: Biol. Sci. 284.

Renault, D., Laparie, M., McCauley, S.J., Bonte, D., 2018. Environmental adaptations, ecological filtering, and dispersal central to insect invasions. Annu. Rev. Entomol. 63, 345–368.

Reusch, T.B.H., Ehlers, A., Hammerli, A., Worm, B., 2005. Ecosystem recovery after climatic extremes enhanced by genotypic diversity. Proc. Natl. Acad. Sci. 102, 2826–2831.

Ricotta, C., Avena, G., 2003. On the relationship between Pielou's evenness and landscape dominance within the context of Hill's diversity profiles. Ecol. Indic. 2 (4), 363–365.

Rius, M., Darling, J.A., 2014. How important is intraspecific genetic admixture to the success of colonising populations? Trends Ecol. Evol. 29, 233–242.

Roff, D., 2001. Life History Evolution. Encyclopedia of Biodiversity. Elsevier, Sunderland, MA, pp. 631–641.

Saastamoinen, M., Bocedi, G., Cote, J., Legrand, D., Guillaume, F., Wheat, C.W., Fronhofer, E.A., Garcia, C., Henry, R., Husby, A., Baguette, M., Bonte, D., Coulon, A., Kokko, H., Matthysen, E., Niitepõld, K., Nonaka, E., Stevens, V.M., Travis, J.M.J., Donohue, K., Bullock, J.M., del Mar Delgado, M., 2018. Genetics of dispersal. Biol. Rev. 6 (93), 574–599.

Schrieber, K., Lachmuth, S., 2017. The Genetic Paradox of Invasions revisited: the potential role of inbreeding × environment interactions in invasion success. Biol. Rev. 92, 939–952.

Shannon, C.E., Weaver, W., 1969. The Mathematical Theory of Communication. The University of Illinois Press, Urbana, IL.

Simberloff, D., 2009. The role of propagule pressure in biological invasions. Annu. Rev. Ecol. Evol. Syst. 40, 81–102.

Sinclair, J.S., Arnott, S.E., Millette, K.L., Cristescu, M.E., 2019. Benefits of increased colonist quantity and genetic diversity for colonization depend on colonist identity. Oikos 128, 1761–1771.

Stauber, L., Badet, T., Feurtey, A., Prospero, S., Croll, D., 2021. Emergence and diversification of a highly invasive chestnut pathogen lineage across southeastern Europe., https://doi.org/10.7554/eLife.56279. https://elifescienes.org/articles/56279.

Stephens, P.A., Sutherland, W.J., Freckleton, R.P., 1999. What is the Allee effect? Oikos 87, 185.

Strubbe, D., Broennimann, O., Chiron, F., Matthysen, E., 2013. Niche conservatism in non-native birds in Europe: niche unfilling rather than niche expansion. Glob. Ecol. Biogeogr. 22, 962–970.

Szűcs, M., Melbourne, B.A., Tuff, T., Hufbauer, R.A., 2014. The roles of demography and genetics in the early stages of colonization. Proc. R. Soc. B Biol. Sci. 281, 20141073.

Szücs, M., Vahsen, M.L., Melbourne, B.A., Hoover, C., Weiss-Lehman, C., Hufbauer, R.A., Schoener, T.W., 2017. Rapid adaptive evolution in novel environments acts as an architect of population range expansion. Proc. Natl. Acad. Sci. U. S. A. 114, 13501–13506.

Taylor, C.M., Hastings, A., 2005. Allee effects in biological invasions. Ecol. Lett. 8, 895–908.

Vahsen, M.L., Shea, K., Hovis, C.L., Teller, B.J., Hufbauer, R.A., 2018. Prior adaptation, diversity, and introduction frequency mediate the positive relationship between propagule pressure and the initial success of founding populations. Biol. Invasions 599 (20), 2451–2459.

Wagner, N.K., Ochocki, B.M., Crawford, K.M., Compagnoni, A., Miller, T.E.X., 2017. Genetic mixture of multiple source populations accelerates invasive range expansion. J. Anim. Ecol. 86, 21–34.

Williams, J.L., Hufbauer, R.A., Miller, T.E.X., 2019. How evolution modifies the variability of range expansion. Trends Ecol. Evol. 34, 903–913.

Further reading

Schlägel, U.E., Grimm, V., Blaum, N., Colangeli, P., Dammhahn, M., Eccard, J.A., Hausmann, S.L., Herde, A., Hofer, H., Joshi, J., Kramer-Schadt, S., Litwin, M., Lozada-Gobilard, S.D., Müller, M.E.H., Müller, T., Nathan, R., Petermann, J.S., Pirhofer-Walzl, K., Radchuk, V., Rillig, M.C., Roeleke, M., Schäfer, M., Scherer, C., Schiro, G., Scholz, C., Teckentrup, L., Tiedemann, R., Ullmann, W., Voigt, C.C., Weithoff, G., Jeltsch, F., 2020. Movement-mediated community assembly and coexistence. Biol. Rev. 95, 1073–1096.

CHAPTER 5

Robust statistical inference for complex computer models

5.1 Introduction

This chapter follows the recent results of Oberpriller et al. (2021), briefly Oberpriller et al. (2021), who, to the present author's knowledge, first formulated and investigated a rigorous statistical theory for robust inference in complex computer ecological simulations. Ecological systems are often complex and interdependent (Levin, 1998). To understand these systems, and to forecast their dynamics under changing conditions, ecologists rely increasingly on complex computer simulations (CCS) which include: process-based models, mechanistic models, and system models; see e.g., Evans et al., 2012; Briscoe et al., 2019; Thompson et al., 2020). The CCS serve, for example, to predict ecosystem responses to climate change (e.g., Cheaib et al., 2012; Rahn et al., 2018). The trend toward an increasing use of CCS mirrors similar developments in other scientific fields, for example, galaxy formation (Somerville and Davé, 2015), macroevolutionary dynamics (Rangel et al., 2018), or epidemiological disease control (Drake et al., 2015).

For any of these models, precise forecasts and correct estimates of predictive uncertainty are paramount, both for their scientific interpretation (Petchey et al., 2015), and for the decision-making and governmental actions (Dietze et al., 2018). The IPCC report (predictiion of climate-change threat) uses for example, a combination of different earth system models to simulate future behavior of the atmosphere, ocean, land surface, and fluxes (Bindoff et al., 2013). Using computer simulations for decision-making is only sensible, however, if their predictions are sufficiently precise, and if their uncertainties are correctly communicated (Budescu et al., 2009).

Achieving these goals depends on correctly determining model structure, parameters, and their uncertainties. Where parameters and model structure cannot be determined directly by measurement or theory, they have to be estimated by comparing model predictions to data (model calibration and selection, e.g., Hartig et al., 2012; Dietze, 2017). In recent years, the field has moved from informal methods for model calibration to established statistical methods such as maximum likelihood estimation (MLE, e.g., Castiglioni et al., 2010) or Bayesian inference (e.g., Harrison et al., 2012; Luke et al., 2017). Superficially, it would seem that

Complexity and Complex Ecological Systems. https://doi.org/10.1016/B978-0-443-19237-1.00003-X

parameter calibration and uncertainty propagation in CCS are not different from the statistical regression models familiar to most ecologists and that no special statistical theory is needed for these models (at least as long as model outputs are "approximately deterministic," for stochastic simulation models, see Hartig et al., 2011).

In practice, however, there are important differences between calibrating simple statistical models and CCS. One trivial difference is the sheer computational challenge of constraining large models to big data (e.g., Fer et al., 2018). Another, more fundamental disparity arises through the model structure. Compared with statistical models, CCS are characterized by having a higher level of interconnectedness and nonlinearity, as well as multiple variables and outputs. Moreover, CCS typically make a large number of structural assumptions based on prior knowledge (Dormann et al., 2012). As a consequence, they are often less flexible in terms of what outputs or patterns can be produced, despite having a large number of parameters (Fatichi et al., 2016).

The above traits lead to certain problems when calibrating those CCS that are less common in statistical models. A particularly important example is trade-offs appearing when calibrating the model to multiple data streams. It has been argued that using multiple data streams is desirable because information from different biological levels of organization (e.g., daily carbon fluxes and yearly inventory data) contains more complementary information than a single data stream (e.g., Grimm, 2005; Medlyn et al., 2015). However, the combination of internal constraints (e.g., mass or energy balance) with structural error will often make it impossible for a CCS to fit all data streams simultaneously (for a list of examples, see MacBean et al., 2016). Moreover, the information or observation density of data at different organizational levels can differ substantially, leading to unbalanced data (substantial differences in the number of observations of different data streams) for the calibration. This means that the calibration cannot avoid a systematic misfit (bias) in some of the model outputs and additionally faces a conflict between the information provided by different, possibly unbalanced data streams, situations, which are less common in statistical models. The goal of this chapter is to explore these problems in more detail, and following the original work by Oberpriller et al. (2021) provides an overview of strategies for robust statistical inference with CCS. In a part of the text, the researchers first explain the problems that may occur when calibrating CCS with structural error, illustrated with the example of a complex forest ecosystem model. Based on the results, the author of the present book adduces the researchers a range of suggested remedies and provides practical recommendations for using statistical inference with CCS in ecology and evolution.

5.2 Why does the model error affect statistics differently?

Oberpriller et al. (2021) state the important question: why does model error affect statistics differently in complex computer simulations (CCS). To start a related discussion, it will be helpful to further clarify how conventional statistical models differ from CCS. Models exist on a continuum between the two classes (Dormann et al., 2012), but considering the ends of this spectrum, Oberpriller et al. (2021) see clear distinctions between models typically used for statistical data analysis (e.g., GLMMs, see Bolker et al., 2009) and CCS (e.g., Trotsiuk et al., 2020). One key difference is that CCS usually connect a sizeable number of state variables via processes that aim to represent our scientific understanding of the natural system, often with submodels that are calculated at different time steps (e.g., daily, weekly, and annual, see as an example the LPJ-GUESS model Smith et al., 2001). It has often been argued that their mechanistic nature makes CCS more appropriate than regression models for forecasting far into the future, because, at least in principle, they should be able to predict into domains for which no previous data exists (e.g., Kearney et al., 2010; Rastetter, 2017; Radchuk et al., 2019).

The benefits of CCS, however, come along with larger structural complexity, which exacerbates challenges in identifying the correct model structure and correcting possible model-data discrepancies (Peng et al., 2011). For example, their typically high interconnectedness hampers the localization of structural errors. Moreover, while their mechanistic underpinning grants better inclusion of prior knowledge regarding the processes driving system dynamics (Dietze et al., 2013a, b), it can become a liability when mechanisms or parameters are unknown and have to be guessed. A final point is that CCS have to apply certain simplifications and discretizations for computational reasons (e.g., discrete soil layers Tiktak and Bouten, 1992). As a result of these and many more challenges, most CCS display certain structural errors, which are difficult to fix immediately (e.g., Richardson et al., 2012).

These structural errors (including observational bias as part of the statistical model) and their associated uncertainties increase the uncertainties in the calibration process (Bayarri et al., 2007; see also Beven, 2005; Trucano et al., 2006). To address this issue, the field has moved toward using formal, statistical methods for model calibration and uncertainty propagation. These methods, however, infer parameters and uncertainties conditional on the assumed model structure being correct. Statistical modelers are usually not overly concerned about these assumptions, because their models flexibly adjust to data, and thus, their main concerns are distributional assumptions (e.g., Warton et al., 2015). In CCS, however, this assumption will not hold, and structural errors will interact with the inference, in particular when nonlinearities are large, and when the model is fit to imbalanced data (Abramowitz et al., 2008), i.e., when one data stream has much more observations than another.

A statistical calibration will respond to this problem by compensating structural error through adjusting parameters to values that differ from the true values of the underlying process (Bell and Schlaepfer, 2016). The resulting model may still display acceptable performance in the domain for which data are available, but parameter estimates may be biased, and their uncertainties may be underestimated. Moreover, when extrapolating beyond the data domain, which is considered an important strength of CCS, biases and underestimation of uncertainty

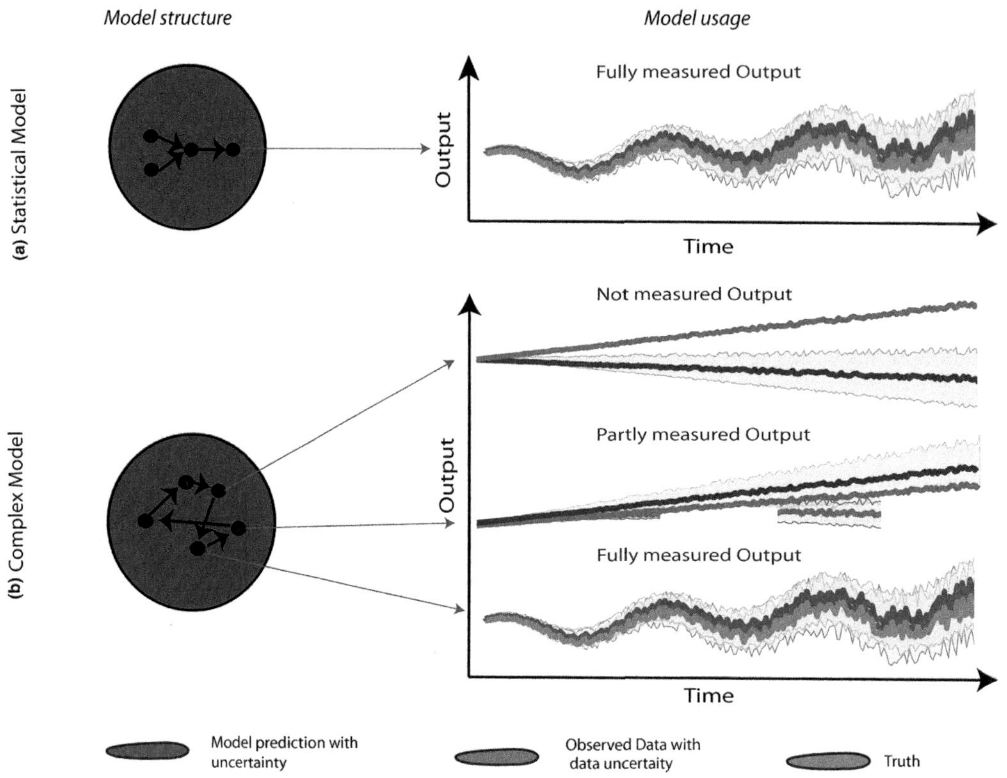

Fig. 5.1

A visualization of differences between complex computer simulations and statistical models. *Based on this figure in Oberpriller, Cameron, Dietze, and Hartig, (2021) Ecology Letters (2022) 24:1251–1261.*

can become substantial (He et al., 2014), especially when the model is calibrated to multiple unbalanced data streams (an example dealing with these issues is Richardson et al., 2010). If the model is not able to fit both data streams at the same time, the calibration algorithm will face a conflict (MacBean et al., 2016). In this situation, the calibration will tend to use parameters adjustments to compensate the error in the more data-rich outputs, at the cost of increased error and too narrow confidence intervals (Sargsyan et al., 2019) particularly in the data-poor model outputs (Fig. 5.1).

While statistical models are generally suited to only one response variable, CCS often predict multiple response variables and thus can serve to multiple data sources, which may vary in sample size and can be used to extrapolate to unobserved variables. Moreover, CCS typically have more variables that are in a more nonlinear and connected dependence structure. From these differences, one may hypothesize that (1) biased complex models will lead to biased parameter estimates and wrong predictions, (2) standard calibration underestimates uncertainty, and (3) both of these problems increase when calibrating against unbalanced data sets.

5.2.1 Case study

To provide a practical example of these problems, the reader can examine the influence of structural model error on predictions, parameters, and uncertainty estimation by calibrating the Basic forest model (BASFOR) authored by Oberpriller et al. (2021) to multiple balanced or unbalanced data streams.

5.2.2 Model structure and introduced structural error

BASFOR simulates horizontal homogeneous forest stands by representing three biogeochemical cycles (carbon, nitrogen, and water) as well as soil environment interaction. It is driven by environmental data (atmospheric CO_2 concentration, solar radiation, air temperature, precipitation, wind speed, and humidity) and describes the forest stand by 17 state variables (nine tree-related and eight soil-related).

To examine the implications of structural error, the reader should modify several key processes in BASFOR. First, he should change the temperature dependence of NPP allocation (higher optimal temperature, fewer allowed deviations). Second, he should make the decomposition of litter temperature dependency. Third, he should change dependence of water runoff to the leaf area index (exponential quadratic instead of exponential linear). Fourthly, he should weight the nitrogen allocation to the tree components with their nitrogen use efficiency. Lastly, he should make nitrogen leaching root-depth dependent. Although the exact location and nature of these modifications are somewhat arbitrary, Oberpriller et al. (2021) regard those modifications as realistic for structural errors that could also occur in real ecosystem models.

5.2.3 Statistical inference

Oberpriller et al. (2021) used the original BASFOR model (henceforth called the "true" model) to simulate synthetic data with random observation errors (0.2) for daily observations of gross primary production (GPP) and daily (balanced data streams) or 10-day (imbalanced data streams, so called because of an unbalance between the number of observations of GPP and ET) measurements of evapotranspiration (ET). Drivers for the simulation were climate data from 1920 to 2005 from Hyytiala, Finland (Reyer et al., 2020).

Prior to the calibration, Oberpriller et al. (2021) conducted a sensitivity analysis of BASFOR. Based on the results, they removed insensitive parameters and three parameters that showed very high trade-offs with other parameters from the calibration by fixing them to their true values (the goal of this procedure is to speed up MCMC computations; see, e.g., Minunno et al., 2013). Because the true parameter values were known, no model error was introduced by this procedure, and the validity of their further results was thus not affected by the parameter screening. In a real application, where "true" parameter values would be unknown, this procedure could introduce additional model error, which would further motivate the need to find methods to compensate for model error, such as the ones presented by Oberpriller et al. (2021).

Oberpriller et al. (2021) applied Bayesian inference (e.g., van Oijen et al., 2011) to infer the values and uncertainties of the remaining six model parameters and the two standard deviation parameters of the observation model from the synthetic data. They specified flat (uniform) priors on the model parameters and vague gamma priors for the standard deviation parameters. They estimated posteriors with the Differential Evolution Markov Chain Monte Carlo (ter Braak and Vrugt, 2008) algorithm, implemented in the R package Bayesian Tools. To speed up computations, Oberpriller et al. (2021) generated initial values and the Z matrix with a differential evolution optimizer DEoptim, (Ardia et al., 2016). They applied this procedure to both the "true" model and the model with structural error.

5.2.4 Quantification of the error in inference

To assess the effect of model error on the inference, Oberpriller et al. (2021) calculated the average error of parameter estimates by averaging the percentage difference between the "true" parameter (p^*) and the calibrated parameter over the posterior, averaged over $N = 10{,}000$ samples from the posterior, the different parameters (P) and the five replicates (M).

$$\text{Parameter error} = P^{-1}\sum\nolimits_i^P\left[M^{-1}\sum\nolimits_J^M N^{-1}\sum\nolimits_k^N \frac{pijk - pi^*}{pi^*}\right] \tag{5.1}$$

Moreover, to assess the error of model predictions (also called time-series error), Oberpriller et al. (2021) calculated the mean absolute error of data d_i and model prediction $m_i(x, \theta_j)$ (driven with climatic drivers x and parameters θ_j) averaged over time (T), the posterior distribution (through $N = 120$ samples from the posterior), and five calibration replicates (M).

$$\text{Error} = M^{-1}\sum\nolimits_J^M N^{-1}\sum\nolimits_k^N T^{-1}\sum\nolimits_i^T \left|d_i - m_i(x,\ \theta_{j,k})\right| \tag{5.2}$$

Note that in most cases with structural model error, the error in the parameters and predictions was systematic, meaning that it can be interpreted as bias.

To relate the error to the estimated uncertainties and thus examine whether uncertainty estimates were reliable, Oberpriller et al. (2021) calculated error scaled to estimated uncertainty (ESEU) by dividing the mean error per day by the posterior standard deviation $\sigma_i(m_i(x,\theta_j))$, averaged over time, the posterior distribution, and the five replicates.

$$\text{ESEU} = \frac{1}{T}\sum\nolimits_i^T \sigma_i^{-1}\left(m_i\left(x,\ \theta_{j,k}\right)\right)\left|d_i - M^{-1}\sum\nolimits_J^M N^{-1}\sum\nolimits_k^N m_i\left(x,\ \theta_j\right)\right| \tag{5.3}$$

A mean absolute error the same magnitude as the estimated uncertainty (standard deviation) will result in an ESEU of 1. Values substantially larger than one suggest that the estimation or prediction error is larger than the estimated uncertainty. For the model outputs and uncertainties, Oberpriller et al. (2021) differentiated between calibration and extrapolation domain.

5.2.5 Comparison between calibrating a "true" model and a model with structural error

The results of the calibration with the "true" model (without structural error) show that the error of the inferred parameters was virtually zero (<0.02%) for balanced and unbalanced data sets (Fig. 5.2A). In both of these cases, extrapolation and calibration error were small with narrow uncertainties (ESEU = 0.1) (Fig. 5.2B).

Performance of the model with and without structural model error for balanced and unbalanced data. The bars reflect error in absolute values, and numbers reflect the error scaled to estimated uncertainty (i.e., the error of the model which can be explained due to a high estimated uncertainty). The case study indicates that structural model bias leads to (a) parameters with serious errors, (b) erroneous model outcomes and high error scaled to estimated uncertainty.

For the model with structural error, inferential errors were much larger (Fig. 5.2A). In particular, the parameter error was three times larger for the unbalanced data (*c.* 5%) compared with the balanced data (*c.* 1.7%) (Fig. 5.2A). Higher parameter error for the model with structural error led in all cases to higher time-series errors compared with the correct model (Fig. 5.2B). For the balanced data set, the error for calibration was smaller than for extrapolation, whereas for the unbalanced data set this only was true for the high-resolution data (GPP). Moreover, GPP error was slightly smaller for the unbalanced than the balanced data set, but ET error otherwise. These errors led to a very high ESEU (Fig. 5.2B). This effect was

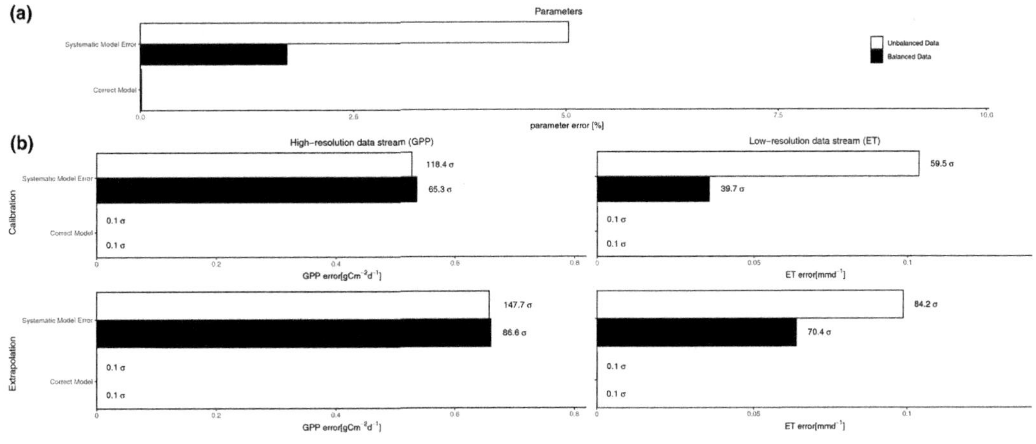

Fig. 5.2

A visualization of differences between complex computer simulations and statistical models. *Based on this figure in Oberpriller, Cameron, Dietze and Hartig, (2021) Ecology Letters (2022) 24:1251–1261.*

stronger for the unbalanced data, especially for the undersampled data (ET) in the calibration domain (Fig. 5.2B).

These results support the theoretical expectations that calibrating with a correct structural model leads to unbiased parameter estimates, correct predictions, and reliable uncertainty estimates, regardless whether data streams are balanced or unbalanced. Introducing structural model error, however, led to erroneous parameter estimations (Fig. 5.2A), caused erroneous time-series predictions and high ESEU (Fig. 5.2B), and these effects were intensified by unbalanced data sets (Fig. 5.2A and B).

5.3 A toolbox for statistical inference in complex computer simulations

After having confirmed the intuition that statistical calibrations of CCS are highly susceptible to structural error, the researchers turned their attention to possible solutions. Few general treatments of the problem exist in literature, but there are certain strategies and suggestions that are frequently used in practice. To deal with the problem of imbalanced data, many studies rebalance or reweight data streams. The remaining model-data discrepancies (bias) have sometimes been addressed by introducing data-driven models to the process-model after or during the calibration. In the following, it is shown how the reader can discuss these potential solutions and test their applicability in the case study of Oberpriller et al. (2021).

5.3.1 Weighting of data streams

The strategy of rebalancing and reweighting data addresses the issue that standard statistical methods weight, the importance of each data stream principally by its content of independent observations. While the latter is perfectly sensible for a correct model, it will lead to distortions toward the model output with more data when structural error makes it impossible to fit both data streams at the same time.

5.3.2 Case study—Weighting of data streams

To examine the possible benefits of weighting for their case study, Oberpriller et al. (2021) down-weighted the likelihood for the GPP data with 1/10, the ratio of ET to GPP observations, thus giving both data streams the same weight. Weighting the data streams increased the error for the estimated parameters of the correct model (Fig. 5.3A) by a small amount, which propagates through the model into a small error in predictions and a higher ESEU (Fig. 5.3B). For the model with structural error, introducing weights in the likelihood decreased parameter error leading to smaller ET error, but slightly increased GPP error (Fig. 5.3B). Moreover, the ESEU of ET in the calibration domain is smaller due to a reduction of ET error. Overall, it can be concluded that weighting slightly decreased the inferential performance for the correct model, but dramatically improved the performance for the model with structural error.

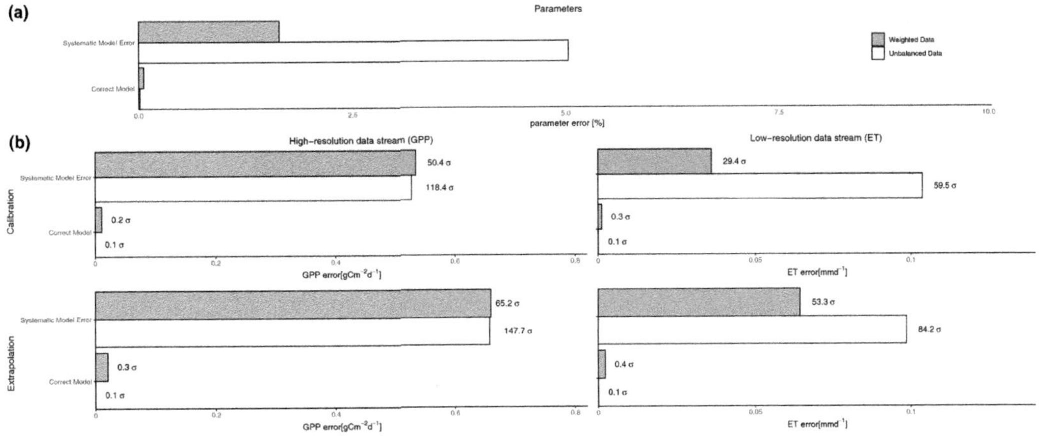

Fig. 5.3

A visualization of differences between complex computer simulations and statistical models. *Based on this figure in Oberpriller, Cameron, Dietze, and Hartig, (2021) Ecology Letters (2022) 24:1251–1261.*

Comparison of the performance of the model with structural model error and the correct model for weighted and unbalanced data. The bars reflect error in absolute values and numbers reflect the error scaled to estimated uncertainty. The case study indicates that weighting the data streams decreases (a) parameters error, (b) shifts error in model outcomes and improves ESEU

5.3.3 Bias correction after calibration

Another option to deal with model error is statistical bias correction. The simplest approach is to fit flexible statistical or machine learning models post hoc (i.e., after the CCS has been calibrated) to the residual errors (but see Beyer et al., 2020). The logic here is that if the model makes the same error under similar conditions (called "time invariance" by Ehret et al., 2012), the error can be identified and corrections can be applied to future predictions. Obviously, this method only corrects predictions and not the parameter estimates, as the actual inference remains unchanged.

Case study—Bias correction after calibration

To test this method, a flexible Gaussian process (GP) model from the kernlab package (Karatzoglou et al., 2004) was used with a distance-based covariance structure (for details see Supporting Information 1, section 1.2). The researchers fitted the model to approximately 6 years of residual errors as a response, and the corresponding model drivers (e.g., temperature and humidity) and CCS output as predictors, and extrapolated the error to future predictions. The results show that this approach decreased the predictive GPP error of the model with structural error by similar amounts in the calibration and extrapolation periods (Fig. 5.4). ET

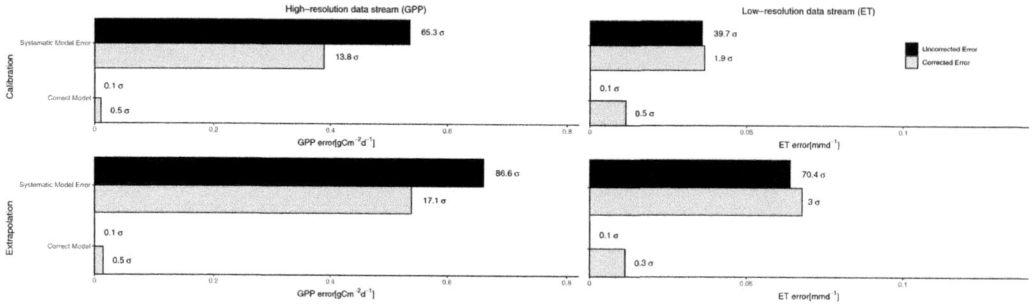

Fig. 5.4

A visualization of differences between complex computer simulations and statistical models. *Based on this figure in Oberpriller, Cameron, Dietze and Hartig, (2021) Ecology Letters (2022) 24:1251–1261.*

error was approximately the same between the corrected and uncorrected versions of the model with a structural error, but there was a large decrease in ESEU (Fig. 5.4), not only caused by reduced error, but mostly by the variance coming from the explicitly modeled model error. Applying the same method to the true model introduced a slightly larger error in the time series and increased ESEU (Fig. 5.5). It may be speculated that this is due to the GP overfitting on random error.

Comparison of the performance of the model with structural error and the correct model fitting a correction term after calibration. The bars reflect error in absolute values and numbers reflect

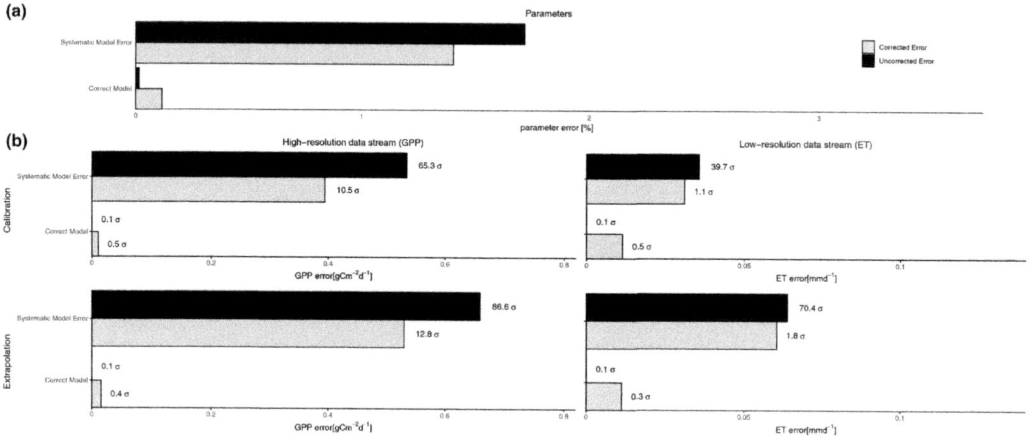

Fig. 5.5

A visualization of differences between complex computer simulations and statistical models. *Based on this figure in Oberpriller, Cameron, Dietze, and Hartig, (2021) Ecology Letters (2022) 24:1251–1261.*

the error scaled to estimated uncertainty. The case study indicates that correcting the data streams decreases error in model outcomes and decreases ESEU.

5.3.4 Bias correction during calibration

Another option is to perform the bias correction within the calibration. A common example of this is the Kennedy-O'Hagan (KOH) approach (Kennedy and O'Hagan, 2001). In this approach, Oberpriller et al. (2021) fit again a GP for the bias together with the other model parameters in the same likelihood.

$$L(\theta) = \left(2^{1/2}\sigma\right)^{-1}\pi^{-1/2}\exp\left\{-2^{-1}\sigma^{-2}[d - (m(\theta, x) + GP(x, m))]^2\right\} \tag{5.4}$$

Here, σ is the standard deviation of the observational error, *GP* a Gaussian process. While the advantage of this approach is that the bias correction can also improve the inference on the model's parameters, the drawback is that it may suffer from an identifiability issue between parameters and model error. Whether this problem occurs depends on how distinct the structure of the process and the error model are. Note also that multiple data streams can be helpful in this regard, because they would typically impose independent constraints on the process model. Moreover, it has been shown that incorporating suitable prior knowledge about the model error (e.g., smooth with respect to some predictor variables) allows the KOH method to separate between parameters and model error (Brynjarsdóttir and O'Hagan, 2014). Because of these attractive properties, there are a sizeable number of studies which have tested and modified this approach (e.g., Higdon et al., 2008; Goldstein and Rougier, 2009; Tuo and Wu, 2016; Tuo, 2017).

Case study—Bias correction during calibration

In its original version, the KOH method fits the GP against all calibration data with all drivers and state variables as predictors. However, as the computational cost of GP fitting and evaluation scale unfavorable with the number of data points, this makes it more difficult for typical environmental model calibrations. The computational problems occur because the calculation of the GP requires an inversion of a large covariance matrix. Moreover, the KOH method assumes having enough observational data of model determining variables (model state and external drivers) to fully constrain the GP (Kennedy and O'Hagan, 2001), which for typical ecological models is not a realistic assumption (in this study, Oberpriller et al. (2021) do not have virtual measurements of any state variables, except the fluxes GPP and ET).

For this case study, Oberpriller et al. (2021) propose an alternative variant of the KOH method, which makes three changes to decrease computational cost. First, they only use the drivers and the observed values as predictors. Secondly, they calibrate against a subsample of data (in their case they subsample to 10% of the data, the last 8 years of data and drivers as best proxies for future drivers). They did so because, typically models systematically predict GPP that is too small on warm summer days and ET that is too high when humidity is low. Thirdly, they avoid the costly inversion of the covariance matrix that is only needed to match GP parameters to their prior by approximating the inverse covariance by its diagonal, while still inferring the full covariance matrix (rbfdot kernel) in the likelihood. To code a preference for explaining the data by the process-model, they apply a regularizing gamma (2,0.1) prior with a high probability weight near zero on the diagonal. Based on the GP predictions, they calculated model-data discrepancies for the rest of the time series.

When applying bias correction during calibration, parameter error stayed near zero for the correct model and decreased for the model with structural error (Fig. 5.5A). However, whereas time series error decreased in both outputs, for the model with structural error, for the true model, error increased (Fig. 5.5B), with an almost identical pattern to the post hoc GP (Fig. 5.5B). For the model with structural error, the calibration resulted in higher estimated uncertainty and thus lower ESEU compared with a calibration without an explicit model error term (Fig. 5.5B). Overall, the method improves parameters, predictions and ESEU for the model with structural error but decreases the performance for the correct model.

Comparison of the performance of the model with the structural model error and the correct model calls for the correction during calibration. The bars reflect error in absolute values, and numbers reflect the uncertainty in units of standard deviation. The case study indicates that correcting error during the calibration decreases (a) parameters error for the wrong model, (b) reduces error in the model outcomes and improves uncertainty estimation.

5.3.5 Correcting processes rather than outputs

Correcting bias on the model outputs can improve predictions and inference. The true error, however, is not on the outputs, but in the model processes themselves. It, therefore, seems obvious to explore if the processes themselves could be bias-corrected. For simple population models, this idea has been suggested under the name "partially specified ecological models" (Wood, 2001). The drawback of this approach for CCS is that the complexity of the error term and therefore the issue of identifiability increases significantly if errors in all possible subprocesses are considered. For their case study Oberpriller et al. (2021) attempted to correct process-errors directly via a state-space approach, but did not succeed in improving the statistical inference in this way. Nevertheless, they believe that such an approach is worthwhile for further research, in particular because it would not only correct errors, but also allow to identify their location.

5.4 Discussion

CCS are increasingly used in ecology, evolution and earth system sciences. Our ability to confront these models with data and to estimate uncertainties in parameters and predictions is critical for their utility.

In their contribution, Oberpriller et al. (2021) highlighted that certain issues emerge when using standard statistical methods to calibrate CCS. Most importantly, their theoretical explanations as well as the case study demonstrated that naive applications of standard calibration methods to imperfect computer simulations can lead to biased parameter estimates and predictions, and to underestimated uncertainties (Fig. 5.2) and that these biases are more pronounced than in flexible statistical models. These issues are particularly severe when calibrating against unbalanced data (Fig. 5.2). Weighting of data streams can reduce the aggravating effect of unbalanced data (Fig. 5.3). Data-driven models can be used to describe and remove the remaining bias after or during the calibration. In their case study, fitting model bias with a GP after calibration improved time series predictions (Fig. 5.4). Thus, the results show that robust methods exist for ameliorating negative consequences of structural model for making predictions with calibrated CCS.

Using a GP during calibration can additionally improve parameter inference (Fig. 5.5). Everyone should acknowledge that the interpretation of parameter values across structurally different models is tricky, because those parameters have different meanings in the respective models, and thus, one could argue that both the true and the model with structural error have parameters that are correct under their respective assumptions. This view, however, neglects, that researchers will tend to interpret parameter values as if their models were structurally unbiased, and representation of the true process. The comparison of the estimated and the true parameter, therefore, measures to what extent this interpretation is justified and shows that explicitly modeling structural error increases the chances of model parameters representing their real values (Goldstein and Rougier, 2009).

The results regarding the consequences of model error are qualitatively supported by the few earlier studies that have looked at the problem (for balanced data by White et al., 2014 and for unbalanced data sets by Abramowitz et al., 2008). In general, however, this topic seems surprisingly underappreciated in the statistical literature. Oberpriller et al. (2021) speculate that most statisticians do not operate with large system models, and the modelers that do are not primarily interested in statistical methods. Nevertheless, a good understanding of these issues is urgently needed, as many important forecasts rely on the correct identification of parameters and their uncertainties. In the next subsections, Oberpriller et al. (2021) summarize their conclusions from existing literature and our new simulations, provide practical guidance for their use, and delineate a statistical research program to develop a theory of robust inference for CCS.

5.4.1 Which methods work to improve inference for biased system models?

To achieve a more balanced impact of the different data on the calibration, many modeling studies weigh data streams. Despite its popularity, few studies have examined the justification for this practice. Contrary to Wutzler and Carvalhais (2014), who only found minor improvements, Oberpriller et al. (2021) found that weighting improved all considered performance measures (Fig. 5.3). Different CCS and a different severity of model error may explain the differences in the two studies. In general, benefits from weighting likely depend on the statistical context, the weighting strategy, and the model error. Overall, however, Oberpriller et al. believe that weighting is a useful and conservative strategy if structure model error is suspected. One open question that would profit from more research is how the weighting of different data streams should be performed. Creating balance by upweighting the less abundant data stream, which essentially corresponds to the common practice of oversampling in machine learning, could lead to a serious underestimation of uncertainties as it is equivalent to using the same data multiple times. Downweighting, the far more common approach in studies calibrating CCS, is more conservative, but it also artificially decreases the information in the more abundant data stream to the level of the less abundant stream, which can hardly be optimal to get realistic uncertainties. In general, these two options represent the extremes of a broad spectrum of possibilities, and more research is required to understand how an optimal weighting could be justified. An option to avoid the problem would be to calibrate against patterns, as suggested by the POM (Grimm, 2005, to independently update subsets of parameters against different data streams (Wutzler and Carvalhais, 2014), or to set up subjective likelihoods (White et al., 2014), as in the GLUE approach (Beven and Binley, 1992). The downside, however, is that these approaches could be considered even more subjective than weights on the data streams.

A complementary class of methods directly addresses the issue of model error, by identifying and correcting structural biases from model's predictions. In this study, the approach (via the KOH method) improves parameters, predictions, and uncertainty quantification (in line with Brynjarsdóttir and O'Hagan, 2014). However, the standard KOH method has two main challenges—high computational complexity (Conti and O'Hagan, 2010) and possible identifiability issues between model parameters and model error (Brynjarsdóttir and O'Hagan, 2014). Oberpriller et al. (2021) addressed the first problem by only using a fraction of the available data to fit the GP and extrapolated to the remaining calibration domain. Oberpriller et al. (2021) speculate subsampling works for models with mechanistic structure, as long as the learned discrepancy will behave similarly in the future. They appreciate that using a fraction of the calibration data potentially disregards useful information and that their additional numerical approximations could further reduce the method's performance. The fact that they reduced the model error, however, suggests that these problems are probably mild. Still, in situations where computational costs are not limiting, it would be better to use the original method suggested in Kennedy and O'Hagan (2001). The issue of identifiability is important, but arises in many statistical situations, and several strategies exist to deal with it, for example regularization or informative priors (Brynjarsdóttir and O'Hagan, 2014) Thus, these methods can lead to better predictions for ecological CCS and modelers should be using them.

A limitation of a case study is that it tests validity and effectiveness for one specific model, with one specific error structure While we do think that the chosen example is typical and representative for the field, it would be useful to explore the generality of the results in future studies and their robustness to observation errors and uncertainties, which can be expected to exacerbate statistical problems.

Finally, all successful examples of Oberpriller et al. (2021) used bias corrections on model outputs. In particular, when making predictions, these implicitly assume that the model error is stationary, which is unlikely to be true (Chen and Brissette, 2015). It would therefore be preferable to move bias corrections directly inside the modeled processes. In their case study, Oberpriller et al. (2021) attempted such a correction with a state-space approach but could not achieve an increase in inferential performance. It is possible that idiosyncrasies of their setup were responsible for this negative result, but it seems equally plausible that corrections on the outputs are already at the limit of what can be sensibly inferred from data. Either way, these considerations suggest that bias corrections are currently no panacea and that careful improvements of the model structure, if possible, are still the preferable solution.

5.4.2 Practical suggestions

As famously noted by Box (1976): “All models are wrong, but some are useful.” Accepting this fact, the question for CCS is what type of error is dominant. If statistical error dominates the structural error (this can be checked by an analysis of residuals, see Supporting information section 2), all standard statistical techniques work fine, regardless of the balance of data. In this case, using methods that accounting for possible structural model errors tends to somewhat increase uncertainties (Figs. 5.4 and 5.5); see also recommendations in Table. 5.1). When structural error dominates, however, severe statistical problems can arise, in particular for imbalanced data. In this case, weighting of data streams or adding bias correction to the CCS can improve the outcomes of a model calibration dramatically. Recommendation of Oberpriller et al. (2021) for modelers with little statistical background is that downweighting imbalanced data is a simple, conservative approach that can alleviate some of issues created by structural error. Although it is somewhat ad hoc, it improved results in the case study, and it makes uncertainty estimates (e.g., confidence intervals) more conservative. For more experienced modelers, Oberpriller et al. (2021) propose to consider additional bias corrections after or during calibration, or even consider if bias corrections can be moved inside the processes, which would not only improve the inference, but also model understanding. For all these purposes, Oberpriller et al. (2021) provide sample code (https://github.com/JohannesOberpriller/Oberpriller-et-al-2021).

The different situations are in environmental model calibration and in Oberpriller et al. (2021) suggestions for improving model performance. The two main factors, which need to be taken into account, are the data situation (balanced or unbalanced) and the sources of error (random or

Table 5.1 Cases, methods, and recommendations according Oberpriller, J., Cameron, D.R., Dietze, M.C., and Hartig, F. (2021).

Cases	Statistical error dominates, balanced or unbalanced data	Structural error dominates, balanced data	Structural error dominates, unbalanced data
Naive use of standard methods	Standard methods are sufficient for parameters, predictions, and uncertainty quantification	Standard methods lead to biased predictions and parameters	Standard methods lead to a higher bias in parameters, predictions, and uncertainty estimation
Our recommendations	1. Standard methods are sufficient	1. Bias correction after calibration improves predictions 2. Bias correction during calibration additionally improves parameters	1. Weighting reduces bias by a lot 2. Bias correction after calibration improves predictions 3. Bias correction during calibration additionally improves parameters
Remarks	-	Bias correction has high computational costs	Bias correction has high computational costs

structural). This general advice can slightly change in different situations as model complexity and computational demand strongly depend on the CCS, domain of extrapolation and number of data streams. Overall performance will become worse with increasing observational error for all methods including standard calibration.

5.4.3 Toward a statistical theory for robust inference in complex computer simulations

More broadly, Oberpriller et al. (2021) highlights that structural model error raises specific problems for statistical inference with CCS. This should alert the ecological community that the model error is a real problem for the calibration of CCS, and naively applying standard statistical methodologies does not always lead to the desired results.

Although a step was made into the direction of robust inference in CCS by reviewing proposed solutions, explaining their theoretical justification and providing practical guidance for their application, further work is required to arrive at a general solution for robust statistical inference. For example, there is no good theory about how to set weights for different data streams. When considering a data stream with only one observation, it becomes clear that downweighting to the least common data stream is likely not always optimal. Moreover, it would be interesting to extend bias corrections also to methods that use simulation-based inference, such as Approximate Bayesian Computing (ABC) or synthetic likelihood (Csilléry et al., 2010; Hartig et al., 2011).

A last point is that statistical bias corrections are important for improving the inference, but the correct model still consistently performed best in the case study of Oberpriller et al. (2021), and we should thus also consider how to develop methods to track down the location of the error. To localize errors, one could start by analyzing model discrepancies for patterns, and use those to attempt a rough localization of the structural error. Moreover, he may speculate that when a dramatic change of a parameter value between KOH and standard calibration happens, this gives a hint that model error affects this specific parameter and thus that model error is "near" to this parameter. Then using time-dependent parameters instead of constant ones, Reichert and Mieleitner (2009), could be an option to identify a better localization of the error. Another idea (Wood, 2001) goes a step further, by saying that the flexible generalized additive models should account for the processes, or by Reichstein et al., 2019), who proposed that the entire submodels should be learned. These approaches should be tested in practice to finally improve model performance, Oberpriller et al. (2021).

References

Abramowitz, G., Leuning, R., Clark, M., Pitman, A., 2008. Evaluating the performance of land surface models. J. Clim. 21, 5468–5481.

Ardia, D., Mullen, K.M., Peterson, B.G., Ulrich, J., 2016. DEoptim: differential evolution in R version 2.24. Available at: https://cratur-project.org./web/packages/DEoptim.

Bayarri, M.J., Berger, J.O., Paulo, R., Sacks, J., Cafeo, J.A., Cavendish, J., et al., 2007. A framework for validation of computer models. Technometrics 49, 138–154.

Bell, D.M., Schlaepfer, D.R., 2016. On the dangers of model complexity without ecological justification in species distribution modeling. Ecol. Model. 330, 50–59.

Beven, K., 2005. On the concept of model structural error. Water Sci. Technol. 52, 167–175.

Beven, K., Binley, A., 1992. The future of distributed models: model calibration and uncertainty prediction. Hydrol. Process. 6, 279–298.

Beyer, R., Krapp, M. & Manica, A. (2020). An empirical evaluation of bias correction methods for palaeoclimate simulations. Clim. Past, 16, 1493–1508. https://doi.org/10.5194/cp-16-1493-2020.

Bindoff, N.L., Stott, P.A., AchutaRao, K.M., Allen, M.R., Gillett, N., Gutzler, D., Hansingo, K., Hegerl, G., Hu, Y., Jain, S., Mokhov, I.I., Overland, J., Perlwitz, J., Sebbari, R., Zhang, X., Plattner, G.-.K., 2013. Detection and attribution of climate change: from global to regional. In: Stocker, T.F., Qin, D., Tignor, M., Allen, S.K., Boschung, J., Nauels, A., Midgley, P.M. (Eds.), Climate Change 2013: The Physical Science Basis. Contribution of Working Group I to the Fifth Assessment Report of the Intergovernmental Panel on Climate Change. Cambridge University Press, Cambridge, UK and New York, NY, USA, pp. 867–952.

Bolker, B.M., Brooks, M.E., Clark, C.J., Geange, S.W., Poulsen, J.R., Stevens, M.H.H., et al., 2009. Generalized linear mixed models: a practical guide for ecology and evolution. Trends Ecol. Evol. 24, 127–135.

Box, G.E.P., 1976. Science and statistics. J. Am. Stat. Assoc. 71, 791–799.

ter Braak, C.J.F., Vrugt, J.A., 2008. Differential evolution Markov Chain with snooker updater and fewer chains. Stat. Comput. 18, 435–446.

Briscoe, N.J., Elith, J., Salguero-Gómez, R., Lahoz-Monfort, J.J., Camac, J.S., Giljohann, K.M., et al., 2019. Forecasting species range dynamics with process-explicit models: matching methods to applications. Ecol. Lett. 22, 1940–1956.

Brynjarsdóttir, J., O'Hagan, A., 2014. Learning about physical parameters: the importance of model discrepancy. Inverse Prob. 30, 114007.

Budescu, D.V., Broomell, S., Por, H.H., 2009. Improving communication of uncertainty in the reports of the intergovernmental panel on climate change. Psychol. Sci. 20, 299–308.

Castiglioni, S., Lombardi, L., Toth, E., Castellarin, A., Montanari, A., 2010. Calibration of rainfall-runoff models in ungauged basins: a regional maximum likelihood approach. Adv. Water Resour. 33, 1235–1242.

Cheaib, A., Badeau, V., Boe, J., Chuine, I., Delire, C., Dufrêne, E., et al., 2012. Climate change impacts on tree ranges: model intercomparison facilitates understanding and quantification of uncertainty. Ecol. Lett. 15, 533–535.

Chen, J., Brissette, F., Lucas-Picher, P., 2015. Assessing the limits of bias correcting climate model outputs for climate change impact studies. J. Geophys. Res. Atmos. 120, 1123–1136.

Conti, S., O'Hagan, A., 2010. Bayesian emulation of complex multi-output and dynamic computer models. J. Stat. Plan. Inference 140, 640–651.

Csilléry, K., Blum, M.G.B., Gaggiotti, O.E., François, O., 2010. Approximate Bayesian Computation (ABC) in practice. Trends Ecol. Evol. 25, 410–418.

Dietze, M.C., 2017. Ecological Forecasting. Princeton University Press, Princeton.

Dietze, M.C., LeBauer, D.S., Kooper, R., 2013a. On improving the communication between models and data. Plant Cell Environ. 36, 1575–1585.

Dietze, M.C., Fox, A., Beck-Johnson, L.M., Betancourt, J.L., Hooten, M.B., Jarnevich, C.S., et al., 2018. Iterative near-term ecological forecasting: needs, opportunities, and challenges. Proc. Natl. Acad. Sci. 115, 1424–1432.

Dietze, M.C., LeBauer, D.S., Kooper, R., 2013b. On improving the communication between models and data. Plant Cell Environ. 36, 1575–1585.

Dormann, C.F., Schymanski, S.J., Cabral, J., Chuine, I., Graham, C., Hartig, F., et al., 2012. Correlation and process in species distribution models: bridging a dichotomy. J. Biogeogr. 39, 2119–2131.

Drake, J.M., Kaul, R.B., Alexander, L.W., O'Regan, S.M., Kramer, A.M., Pulliam, J.T., et al., 2015. Ebola cases and health system demand in liberia. PLoS Biol. 13, e1002056.

Ehret, U., Zehe, E., Wulfmeyer, V., Warrach-Sagi, K., Liebert, J., 2012. Should we apply bias correction to global and regional climate model data? Hydrol. Earth Syst. Sci. Discuss. 9, 5355–5387.

Evans, M.R., Norris, K.J., Benton, T.G., 2012. Predictive ecology: systems approaches. Philos. Trans. R. Soc. B: Biol. Sci. 367, 163–169.

Fatichi, S., Vivoni, E.R., Ogden, F.L., Ivanov, V.Y., Mirus, B., Gochis, D., et al., 2016. An overview of current applications, challenges, and future trends in distributed process-based models in hydrology. J. Hydrol. 537, 45–60.

Fer, I., Kelly, R., Moorcroft, P.R., Richardson, A.D., Cowdery, E.M., Dietze, M.C., 2018. Linking big models to big data: efficient ecosystem model calibration through Bayesian model emulation. Biogeosciences 15, 5801–5830.

Goldstein, M., Rougier, J., 2009. Reified Bayesian modelling and inference for physical systems. J. Stat. Plan. Inference 139, 1221–1239.

Grimm, V., 2005. Pattern-oriented modeling of agent-based complex systems: lessons from ecology. Science 310, 987–991.

Harrison, K.W., Kumar, S.V., Peters-Lidard, C.D., Santanello, J.A., 2012. Quantifying the change in soil moisture modeling uncertainty from remote sensing observations using Bayesian inference techniques. Water Resour. Res. 48. https://doi.org/10.1029/2012wr012337.

Hartig, F., Calabrese, J.M., Reineking, B., Wiegand, T., Huth, A., 2011. Statistical inference for stochastic simulation models—theory and application. Ecol. Lett. 14, 816–827.

Hartig, F., Dyke, J., Hickler, T., Higgins, S.I., O'Hara, R.B., Scheiter, S., et al., 2012. Connecting dynamic vegetation models to data—an inverse perspective. J. Biogeogr. 39, 2240–2252.

He, Y., Yang, J., Zhuang, Q., McGuire, A.D., Zhu, Q., Liu, Y., et al., 2014. Uncertainty in the fate of soil organic carbon: a comparison of three conceptually different decomposition models at a larch plantation. J. Geophys. Res. Biogeosci. 119, 1892–1905.

Higdon, D., Gattiker, J., Williams, B., Rightley, M., 2008. Computer model calibration using high-dimensional output. J. Am. Stat. Assoc. 103, 570–583.

Karatzoglou, A., Smola, A., Hornik, K., Zeileis, A., 2004. Kernlab—an S4 Package for kernel methods in R. J. Stat. Softw. 11, 1–20.

Kearney, M.R., Wintle, B.A., Porter, W.P., 2010. Correlative and mechanistic models of species distribution provide congruent forecasts under climate change. Conserv. Lett. 3, 203–213.

Kennedy, M.C., O'Hagan, A., 2001. Bayesian calibration of computer models. J. R. Stat. Soc. Ser. B (Stat. Methodol.) 63, 425–464.

Levin, S.A., 1998. Ecosystems and the biosphere as complex adaptive systems. Ecosystems 1, 431–436.

Luke, A., Vrugt, J.A., AghaKouchak, A., Matthew, R., Sanders, B.F., 2017. Predicting nonstationary flood frequencies: evidence supports an updated stationarity thesis in the United States. Water Resour. Res. 53, 5469–5494.

MacBean, N., Peylin, P., Chevallier, F., Scholze, M., Schürmann, G., 2016. Consistent assimilation of multiple data streams in a carbon cycle data assimilation system. Geosci. Model Dev. 9, 3569–3588.

Medlyn, B.E., Zaehle, S., De Kauwe, M.G., Walker, A.P., Dietze, M.C., Hanson, P.J., et al., 2015. Using ecosystem experiments to improve vegetation models. Nat. Clim. Chang. 5, 528–534.

Minunno, F., van Oijen, M., Cameron, D.R., Pereira, J.S., 2013. Selecting parameters for bayesian calibration of a process-based model: a methodology based on canonical correlation analysis. SIAM/ASA J. Uncert. Quant. 1, 370–385.

Oberpriller, J., Cameron, D.R., Dietze, M.C., Hartig, F., 2021. Towards robust statistical inference for complex computer models. Ecol. Lett. https://doi.org/10.1111/ele.13728.

van Oijen, M., Cameron, D.R., Butterbach-Bahl, K., Farahbakhshazad, N., Jansson, P.E., Kiese, R., et al., 2011. A Bayesian framework for model calibration, comparison and analysis: application to four models for the biogeochemistry of a Norway spruce forest. Agric. For. Meteorol. 151, 1609–1621.

Peng, C., Guiot, J., Wu, H., Jiang, H., Luo, Y., 2011. Integrating models with data in ecology and palaeoecology: advances towards a model—data fusion approach. Ecol. Lett. 14, 522–536.

Petchey, O.L., Pontarp, M., Massie, T.M., Kéfi, S., Ozgul, A., Weilenmann, M., et al., 2015. The ecological forecast horizon, and examples of its uses and determinants. Ecol. Lett. 18, 597–611.

Radchuk, V., Kramer-Schadt, S., Grimm, V., 2019. Transferability of mechanistic ecological models is about emergence. Trends Ecol. Evol. 34, 487–488.

Rahn, E., Vaast, P., Läderach, P., van Asten, P., Jassogne, L., Ghazoul, J., 2018. Exploring adaptation strategies of coffee production to climate change using a process-based model. Ecol. Model. 371, 76–89.

Rangel, T.F., Edwards, N.R., Holden, P.B., Diniz-Filho, J.A.F., Gosling, W.D., Coelho, M.T.P., et al., 2018. Modeling the ecology and evolution of biodiversity: biogeographical cradles, museums, and graves. Science 361, eaar5452.

Rastetter, E.B., 2017. Modeling for understanding v. Modeling for numbers. Ecosystems 20, 215–221.

Reichert, P., Mieleitner, J., 2009. Analyzing input and structural uncertainty of nonlinear dynamic models with stochastic, time-dependent parameters. Water Resour. Res. 45. https://doi.org/10.1029/2009wr007814.

Reichstein, M., Camps-Valls, G., Stevens, B., Jung, M., Denzler, J., Carvalhais, N., et al., 2019. Deep learning and process understanding for data-driven Earth system science. Nature 566, 195–204.

Reyer, C.P.O., Silveyra Gonzalez, R., Dolos, K., Hartig, F., Hauf, Y., Noack, M., et al., 2020. The PROFOUND Database for evaluating vegetation models and simulating climate impacts on European forests. Earth Syst. Sci. Data 12, 1295–1320.

Richardson, A.D., Anderson, R.S., Arain, M.A., Barr, A.G., Bohrer, G., Chen, G., et al., 2012. Terrestrial biosphere models need better representation of vegetation phenology: results from the North American carbon program site synthesis. Glob. Chang. Biol. 18, 566–584.

Richardson, A.D., Williams, M., Hollinger, D.Y., Moore, D.J.P., Dail, D.B., Davidson, E.A., et al., 2010. Estimating parameters of a forest ecosystem C model with measurements of stocks and fluxes as joint constraints. Oecologia 164, 25–40.

Sargsyan, K., Huan, X., Najm, H., 2019. Embedded model error representation for bayesian model calibration. Int. J. Uncertain. Quantif. 9, 365–394.

Smith, B., Prentice, I.C., Sykes, M.T., 2001. Representation of vegetation dynamics in the modelling of terrestrial ecosystems: comparing two contrasting approaches within European climate space. Glob. Ecol. Biogeogr. 10, 621–637.

Somerville, R.S., Davé, R., 2015. Physical models of galaxy formation in a cosmological framework. Annu. Rev. Astron. Astrophys. 53, 51–113.

Thompson, P.L., Guzman, L.M., Meester, L.D., Horváth, Z., Ptacnik, R., Vanschoenwinkel, B., et al., 2020. A process-based metacommunity framework linking local and regional scale community ecology. Ecol. Lett. 23 (9), 1314–1329.

Tiktak, A., Bouten, W., 1992. Modelling soil water dynamics in a forested ecosystem. III. Model description and evaluation of discretization. Hydrol. Process. 6, 455–465.

Trotsiuk, V., Hartig, F., Cailleret, M., Babst, F., Forrester, D.I., Baltensweiler, A., et al., 2020. Assessing the response of forest productivity to climate extremes in Switzerland using model—data fusion. Glob. Chang. Biol. 26, 2463–2476.

Trucano, T.G., Swiler, L.P., Igusa, T., Oberkampf, W.L., Pilch, M., 2006. Calibration, validation, and sensitivity analysis: What's what. Reliab. Eng. Syst. Saf. 91, 1331–1357.

Tuo, R., 2017. Adjustments to Computer Models via Projected Kernel Calibration. arXiv:1705.03422 [stat].

Tuo, R., Wu, C.F.J., 2016. A theoretical framework for calibration in computer models: parametrization, estimation and convergence properties. SIAM/ASA J. Uncert. Quant. 4, 767–795.

Warton, D.I., Foster, S.D., De'ath, G., Stoklosa, J., Dunstan, P.K., 2015. Model-based thinking for community ecology. Plant Ecol. 216, 669–682.

White, J.T., Doherty, J.E., Hughes, J.D., 2014. Quantifying the predictive consequences of model error with linear subspace analysis. Water Resour. Res. 50, 1152–1173.

Wood, S.N., 2001. Partially specified ecological models. Ecol. Monogr. 71, 1–25.

Wutzler, T., Carvalhais, N., 2014. Balancing multiple constraints in model-data integration: weights and the parameter block approach. J. Geophys. Res. Biogeosci. 119, 2112–2129.

CHAPTER 6

Biodiversity maintenance in food webs

6.1 Introduction

This chapter treats biodiversity maintenance in food webs supported mainly (most often) by regulatory environmental feedbacks. A food web consists of all the food chains in a single ecosystem. Each living thing in an ecosystem is part of multiple food chains. Each food chain is one possible path that energy and nutrients may take as they move through the ecosystem. Although the food web is one of the most fundamental and oldest concepts in ecology, elucidating the strategies and structures by which natural communities of species persist remains a challenge to empirical and theoretical ecologists. Consistently with the current literature (e.g., Bagdassarian et al., 2007), it is shown that simple regulatory feedbacks between autotrophs and their environment when embedded within complex and realistic food-web models enhance biodiversity. The food webs are generated through the niche-model algorithm and coupled with predator-prey dynamics, with and without environmental feedbacks at the autotroph level. With high probability and especially at lower, more realistic connectance levels, regulatory environmental feedbacks result in fewer species extinctions, that is, in increased species persistence. These same feedback couplings, however, also sensitize food webs to environmental stresses leading to abrupt collapses in biodiversity with increased forcing. Feedback interactions between species and their material environments anchor food-web persistence, adding another dimension to biodiversity conservation. Therefore, it is suggested that the regulatory features of two natural systems, deep-sea tubeworms with their microbial consortia and a soil ecosystem manifesting adaptive homeostatic changes, can be embedded within niche-model food-web dynamics.

Organisms within an ecosystem are constantly interacting with and altering their abiotic environment. While classic food-web studies have focused mainly on trophic interactions and energy fluxes (Elton, 1927; Lindeman, 1942), unraveling the effects of environmental feedbacks and environmental modification by organisms on biodiversity maintenance is a current focus in ecological research. Current stability-complexity studies continue to be based upon model food webs with diverse predator-prey links or number of species. There predator-prey dynamics are described by increasingly sophisticated systems of differential equations substantiated by Lotka-Volterra-like formulations (Bagdassarian et al., 2007). However, these systems are as a rule prone to meaningful numbers of species extinctions. Basic understanding of the mechanisms along with the identification of the mechanisms ensuring

Complexity and Complex Ecological Systems. https://doi.org/10.1016/B978-0-443-19237-1.00009-0

food-web integrity remains a challenge, and several schemes leading to increased species persistence have been proposed.

In Elsevier's volume on the classification of estuarine and nearshore coastal ecosystems, Simenstad and Yanagi (2011, b) describe the diverse approaches that scientists and managers have taken to classify estuaries and nearshore coasts. Classification of estuaries and coasts is motivated by different perspectives (e.g., landform geomorphology, evolutionary origins, and formative processes), purposes (e.g., understanding structure, variability and dynamics, functions and values, and interaction with adjoining fluvial and coastal ecosystems), and applications (e.g., categorizing, mapping, and management) over diverse temporal and spatial scales. As experienced throughout the history of most sciences, classification of diverse objects is a foundational step in the progress and application of a discipline, and particularly so in applied sciences. Appreciating how different classes of estuaries and coasts evolve and function is a prerequisite to identifying the approaches and tools needed to management issues and drivers and to identify and predict change and assess impacts. However, as useful as these disciplinary classifications are, interdisciplinary classifications remain elusive, especially those linking the geomorphic form and physical structure and dynamics with ecological and water quality or broader ecosystem functions, goods, and services.

Figs. 6.1–6.4, i.e., A, B, C and D, describe results for 10-species food beds and other characteristics explained in the captions.

Ecological models show that complexity usually destabilizes food webs, predicting that they should not amass large numbers of interacting species that are in fact found in nature. Applying nonlinear models, one may study the influence of interaction strength (likelihood of consumption of one species by another) on the food-web dynamics away from equilibrium. Consistent results show that weak-to-intermediate strength links are important in promoting community persistence and stability. Weak links dampened oscillations between consumers and resources. All this tends to maintain population densities away from zero, decreasing the

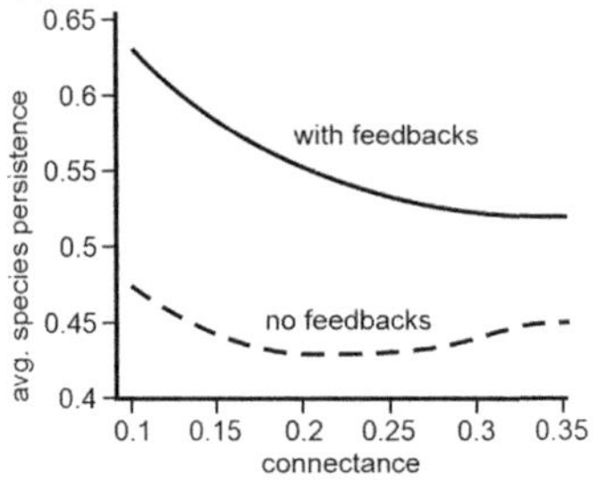

Fig. 6.1
Results for 10-species food webs. Environmental feedbacks lead to increased average species persistence through the entire connectance range.

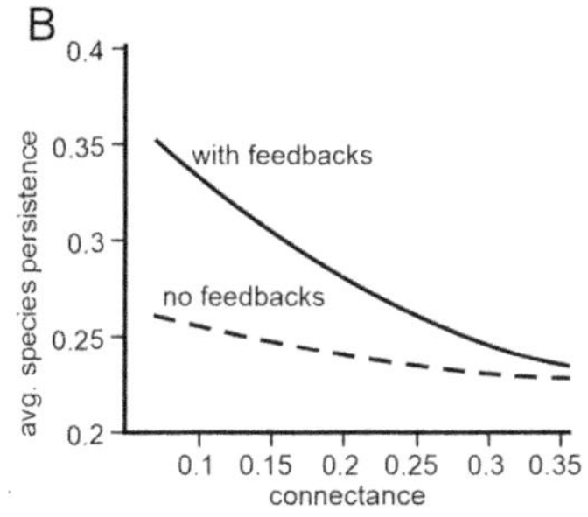

Fig. 6.2

Results for 20-species food webs. For both 10- and 20-species webs, the mass-specific metabolic rate $x_i = 0.3$ for autotrophs and 0.2 otherwise, the maximum rate at which species *i* consumes *j* is $y_{ij} = 3.5$, luminosity is tuned to 0.94, and DW coupling $\gamma = 0.00326$. Moderate deviations from the luminosity value (± 0.06) do not affect the results. Polynomial curve fit to data sets.

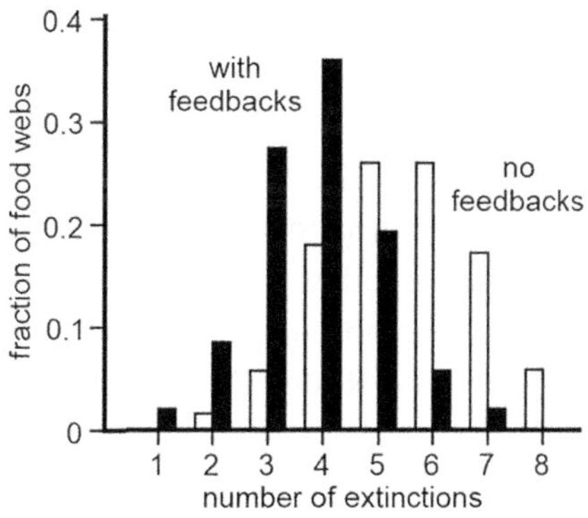

Fig. 6.3

With environmental feedbacks, the distribution of species extinctions shifts to the left to peak at smaller numbers. The maximum number of extinctions for feedback-regulated webs is 7 species, while the minimum number in the unregulated case is found to be 2. Results are for 10-species food webs with connectance $C = 0.12$ and with parameters as in Fig. 6.2. Feedback scenario represented by black-fill bars; no feedback scenario represented by white-fill bars.

statistical chance that a population will become extinct (lower population densities are more prone to such chances). Suitable data on interaction strengths in the natural food webs indicate that food-web interaction strengths are indeed characterized by many weak interactions and a few strong interactions.

Encyclopedia of Soils in the Environment, Hillel (2004), is a valuable encyclopedic resource about soil rich mix of mineral particles, organic matter, gases, and soluble compounds that foster both plant and animal growth. Civilization depends more on the soil as human populations continue to grow, and increasing demands are placed upon available resources. The

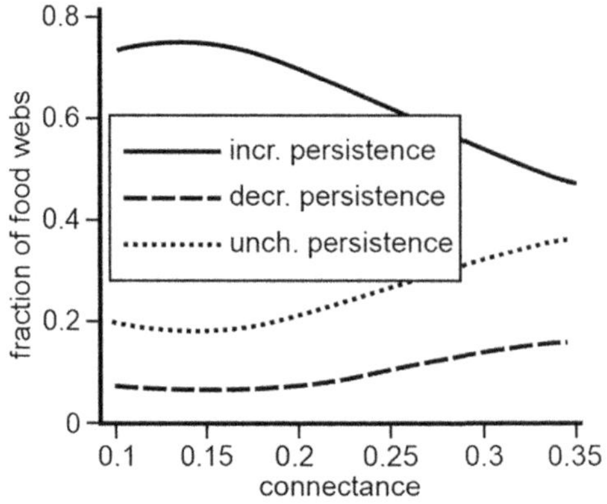

Fig. 6.4
Fraction of food webs showing increased, decreased, or unchanged persistence when environmental feedbacks are on. Results are for 10-species food webs with parameters as in Fig. 6.2.

Encyclopedia is a comprehensive and integrated consideration of the topic of vital importance to human societies in the past, present, and future. This collection encompasses the present knowledge on the world's variegated soils, their origins, properties, classification, and roles in the biosphere. A team of well-known international contributors have written over 250 entries that cover a broad range of issues facing today's soil scientists, ecologists, and environmental scientists. This four volume set features thorough articles that survey specific aspects of soil biology, ecology, chemistry, and physics. Rounding out the encyclopedia's coverage, contributions cover cross-disciplinary subjects, such as the history of soil utilization for agricultural and engineering purposes and soils in relation to the remediation of pollution and the mitigation of global climate change.

Also available is the online navigation via Elsevier's ScienceDirect featuring extensive browsing, searching, and internal cross-referencing articles in the work, plus dynamic linking to journal articles and abstract databases. For more information and availability, visit www.info.sciencedirect.com. Organized encyclopedic format provides concise, readable entries, easy searches, and cross-references. Abundant visual resources: photographs, figures, tables, and graphs are present in every entry. All this provides a broad up-to-date coverage of many important topics, essential information for scientists, students, and professionals.

We stress first the role of weak trophic interactions and the balance of nature. Trophic interactions are likely to affect the distribution and abundance of organisms in fundamental ways, since the success of populations is largely a function of benefits derived from the acquisition of energy (and nutrients) and losses derived from the predation. Food-web descriptions of the soil community provide a way to analyze the dynamics and persistence of the various populations in the context of stability of the community as a whole. The interaction strengths are central in analyses of the trophic interactions in community stability. Interaction strengths refer to the per capita—in this case per biomass—mutual effects. These can be derived from the population sizes and energy flow rates (i.e., the feeding rates) by assuming Lotka-Volterra equations for the dynamics of the functional groups:

$$dX_i/dt = X_i\left(b_i + \Sigma n_j \ln c_{ij}X_j\right) \tag{6.1}$$

where X_i and X_j represent the population sizes of group i and j, respectively; b_i is specific rate of increase or decrease of group i; and c_{ij} is the coefficient of interaction between group i and group j. Mathematically, interaction strengths are defined as entries of the Jacobian community matrix (a_{ij}) being the partial derivatives near the equilibrium: $\alpha_{ij} = (\partial X_i / \partial X_j)^* = (\delta d_t X_i / \delta X_j)^*$. Values of interaction strengths can be derived by equating the death rate of a group i due to predation by a group j in the equilibrium, $c_{ij}X_i^*X_j^*$, to the mean annual feeding rate, F_{ij}, and the production rate of group j due to feeding on group i, $c_{ji}X_j^*X_i^*$, to $a_j p_j F_{ij}$. With the equilibrium population sizes, X_i^*, X_j^*, assumed to be equal to the observed annual mean population sizes, B_i, B_j, the effect of predator j on prey i is

$$\alpha_{ij} = c_{ij}X_i^* = -F_{ij}/B_j \tag{6.2}$$

and the effect of prey i on predator j is:

$$\alpha_{ij} = c_{ji}X_j^* = a_j p_j F_{ij}/B_i \tag{6.3}$$

Estimates of the interaction strengths obtained this way for the soil food webs reveal patterns along the trophic position, characterized by relatively strong top-down effects at the lower trophic levels and relatively strong bottom-up effects at the higher trophic levels, see for instance, Fig. 2, in Encyclopedia of Soils in the Environment, (Hillel, 2004). The patterns of interaction strengths are important to the community stability, as indicated by the comparison between the stability of community matrix representations of seven soil food webs (from the prairie and arable soils). This stemmed from the empirically based values of interaction strengths ("real" matrices) and matrices in which these values were randomized. The comparison shows that the matrices including the realistic patterns of interaction strengths have a much higher level of stability than their randomized counterparts. Graphical interpretation shows effects of the patterning of interaction strengths on the stability of seven soil food webs from prairie and arable land. The black fraction in the bars denotes the level of stability (percentage) based on thousand model runs. Stability of the community matrices is established by evaluating the signs of eigenvalues of the matrices; when all real parts are negative, the matrix is stable and the food web is considered to be locally stable. The stabilizing patterns of the interaction strengths are the direct result of patterns in the energetic properties of the food webs such as the population sizes (biomasses) and feeding rates, Eqs. (6.2), (6.3). Therefore, whenever the soil biology is looked at in terms of the trophic interactions in the soil food web, the structure of the community and the dynamics of the soil populations become inextricably interrelated with soil ecosystem processes and functioning.

6.2 Toward trophic relationships of coastal and estuarine ecosystems

Asmus and Asmus (2011), consider the general aspects of suspension-feeder communities. Food webs of suspension-feeder communities in intertidal and sub-tidal areas yield a general portrait of the dominant components of these communities. A detailed case study of an intertidal mussel bed is performed by using the results of network analysis. Comparisons are made between North Sea mussel beds and oyster and mussel beds at the Atlantic Coast of France and the East Coast of North America. The roles of suspension-feeder communities in the food webs of coastal areas are meaningful. Trophic interactions between these communities and their ambient environment are among the most important ecological processes in shallow waters and may dominate the benthic-pelagic coupling particularly in coastal areas (Prins and Smaal, 1990a, b; Dame et al., 1991a, b; Smaal and Haas, 1997; Smaal and Zurburg, 1997; Prins et al., 1998; Asmus and Asmus, 2005). They state that there have only been a few attempts to quantify the trophic web of suspension-feeder communities integrating the different pathways from grazing phytoplankton to predation by birds.

In the work by Baird and Suthers (2007), a size-resolved pelagic ecosystem model is developed using descriptions of physical limits to biological processes and allometric relationships to determine physiological rates. The model contains three functional groups: phytoplankton, protozoans, and metazoans, all requiring three separately resolved size distributions. Within each functional group, the size-resolution of the model can be altered without changing the model parameters, i.e., the coefficients of allometric relationships, or with changing the model equations, which are characteristic of each functional group. This approach allows the number of size-classes to be varied, and for a convergence of output with increasing resolution to be achieved (Baird and Suthers, 2007).

Many studies on the structure and function of marine and coastal ecosystems based on the analysis of quantitative food webs have been published. Most of these use network analysis, a set of algorithms derived from input-output analysis, trophic and cycle analysis, and the information theory-based computation of system level properties. This sort of the information reflects the complexity of organization in the system. Some of the parameters are: (1) TST, total system throughput or the sum of all flows through the system and some indication of the activity of the system; (2) A, ascendency, a measure of how well a system is performing and incorporates both the size and organization of flows (ascendency is the product of TST and the average mutual information inherent in the network); (3) DC, the upper limit of ascendency called the development capacity, in fact the product of TST and the flow diversity; (4) θ, system overhead, numerically represented by the difference of DC-A, i.e., the cost to the system to operate in a standard way; and (5) flow diversity, defined as DC/TST, which encompasses both the numbers and of interactions and evenness of flows in the food-web network. Results from the analysis of flow networks of ecosystems provide the information that can be used to approach environmental problems at the level of the whole ecosystem and can fruitfully be used to compare ecosystems on spatial and temporal scales, Baird et al. (2007). The intertidal area of the Sylt-Rømø Bight was divided into 8 benthic and 1 pelagic subsystems according to habitat nature and the unique biodiversity of each. A quantitative food-web network was constructed for each of the subsystems. Each flow model comprised 56 living and 3 nonliving compartments and depicted the biomass of each, a balanced energy budget for each of the living components, and the flow of energy and material between all compartments. These models were analyzed by means of the network analysis that revealed a considerable amount of variability between them in terms of system properties such as TST, development capacity, ascendency and redundancy, and a number of dimensionless ratios used in comparative systems of ecology. Mussel beds stood out as the most productive subsystems at $5095 mgCm^{-2} d^{-1}$, with a high TST of $33{,}571 mgCm^{-2} d^{-1}$. The amount of material recycled in each system, ranged from a high of 28% in the muddy sand flats to a low of 2.5% in the mussel beds, while the efficiency of energy transfer in the various systems fluctuated from a low of 3.3% in the sandy shoals to a high of 15% in the mussel beds. Mussel beds were highly specialized in terms of ascendancy and average mutual information in comparison with the other subsystems, but had less resilience. For the two mutually exclusive system attributes, ascendency and redundancy, most of the systems showed ratios between 0.8 and 1.4. Relative redundancy indices calculated for the pelagic and mussel bed subsystems were low, indicating less organized systems with less resistance to disturbances.

In the intertidal area, the effect of suspension-feeder communities is most pronounced, because the water column is shallow and mixing of tidal water is intense; thus, suspension feeders may be able to use the total water column for feeding (Asmus et al., 1992; Asmus and Asmus, 2000). Where suspension feeders such as mussels occur on soft bottoms, intense filtering, feeding, and digestive processes lead to a high production of feces accumulating among and beyond the mussels and creating organic-rich sediments, which are suitable places for bacterial decomposition and detritivorous infauna (Commito and Boncavage, 1989; Commito et al., 2008).

Shell-bearing suspension feeders, such as mussels and oysters, are also suitable substrates for other hard-bottom flora and fauna characterizing suspension-feeder communities, which can be considered as oases of hard-bottom dwellers in a sandy or muddy surrounding. The large aggregation of biomass of the suspension feeders in these communities attracts many invertebrate and vertebrate predators. Estimates of the interaction strengths for the soil food webs reveal patterns along trophic position, characterized by relatively strong top-down effects at the lower trophic levels and relatively strong bottom-up effects at the higher trophic levels. The patterns of interaction strengths are important to the community stability as is indicated by a comparison between the stability of community matrix representations of seven soil food webs (from the prairie and arable soils), using the empirically based values of interaction strengths ("real"matrices) and matrices in which these values are randomized. The comparison shows that the matrices including realistic patterns of interaction strengths have a much higher level of stability than their randomized counterparts.

By studying various figures, one recognizes effects of the patterning of interaction strengths on the stability of seven soil food webs from prairie and arable land. The black fraction in the bars denotes the level of stability (percentage) based on thousand model runs. Stability of the community matrices is established by evaluating the signs of eigenvalues of the matrices; when all real parts are negative, the matrix is stable and the food web is considered to be locally stable. The stabilizing patterns of the interaction strengths are the direct result of patterns in the "energetic" properties of the food webs such as the population sizes (biomasses) and feeding rates. Therefore, when the soil biology is looked at in terms of trophic interactions in the soil food web, the structure of the community and the dynamics of the soil populations become inextricably interrelated with soil ecosystem processes and its functioning. The large aggregation of biomass of the suspension feeders in these communities attracts many invertebrate and vertebrate predators.

The role of suspension feeders varies according to the spatial and temporal scales at which assessments are made. Indicators of biodiversity, productivity, and filtration capacity are used to assess the influences of suspension feeders at scales ranging from individual mussels to biogeographic regions. Quantitative comparisons based on the "catchment areas" of single mussels, mussel beds, and entire bays are used to illustrate how the role of these organisms varies as a function of the unit of measurement. One key factor influencing the relative importance of suspension feeders at different scales is the rate of water movement and, thus, the volume of water available to the consumers. The biodiversity within the suspension-feeder guilds is important because of the way it can affect the amount and sizes of particles removed from the spectrum of available food items. Variations in the role of the animals are also observed at time scales. The complexity of the scaling problem is illustrated using examples from the suspension-feeder guild in a tidal basin of the Wadden Sea (North Sea) where experiments and field measurements have provided insights into processes and mechanisms accounting for spatial and temporal variations.

According to Smaal and Haas (1997), analysis of food-web structure and temporal dynamics is essential to understanding energy flow and population dynamics of species and may contribute to conservation, wildlife management, and disease and pest control. Their report synthesizes all the observational studies of food-web dynamics to which people have access. Most published food webs are static and cumulative: they depict information gathered over many occasions. A web observed over a single, relatively short time period is time specific. Here, the researchers analyze the relation between cumulative and time-specific versions of webs in 16 published cases. Fourteen of the 16 webs are from detritus-based habitats that harbor large fractions of arthropod species: carcasses, tree fluxes, felled logs, tree holes, dung pads, and an acidic pond. The other two webs describe soybean fields and the arctic tundra. These webs are presented in a consistent format and are analyzed in four ways.

First, Smaal and Haas (1997) quantify temporal trends and levels of variation in nine web properties: the percentages of species in the web that are top predators ($\%T$), intermediate species ($\%I$), and basal species ($\%B$); the ratio of number of prey species to number of predator species (P); the mean chain length (μ); the product of species richness and connectance ($S \times C$);

and the numbers of total species, newly arriving species, and local extinctions. In most webs, $\%I$ and $\%T$ fluctuated widely; the latter generally increased in time or remained constant, while the former correspondingly decreased or remained constant. Since the number of basal species usually varied little, changes in $\%B$ were inversely associated with changes in species richness over successional and seasonal time scales. Basal species occupy the lowest trophic level as the primary producer. They convert inorganic and chemical energy and use solar energy to generate chemical energy. The second trophic level consists of herbivores. The remaining trophic levels include carnivores that consume animals at trophic levels below them. Predictable changes in P, μ and $S \times C$ accompanied the changes in $\%B$, $\%I$, and $\%T$. The numbers of total species, new arrivals, and local extinctions displayed no consistent increasing or decreasing trends.

Second, Smaal and Haas (1997) compare cumulative and time-specific webs from the same habitat to determine which properties, if any, of time-specific webs might be predicted from cumulative webs. In cumulative webs, P, μ, and $\%T$ came closest to the median of the values from time-specific webs, followed by $\%I$, $S \times C$, and $\%B$ cumulative webs, which usually appear in general ecology textbooks, overestimate $S \times C$ and underestimate $\%B$ relative to time-specific versions. In five studies, cumulative webs were completed when the last or next-to-last samples were taken; additional sampling in these cases would probably have uncovered more species.

Third, Smaal and Haas (1997) remove opportunistic species from four time-specific webs to determine how these species influenced web structure. Removing one top-feeding opportunistic species from each web caused a dramatic rise in $\%T$, small reductions in $\% I$, $S \times C$, μ, and P, and a negligible rise in $\%B$. A single opportunistic species, even though it makes only rare and brief visits to a habitat, can dramatically reshape web structure.

Fourth, Smaal and Haas (1997) compare properties of cumulative and time-specific versions of the 16 food webs to properties of cumulative webs in two published web catalogs. The cumulative versions of the 16 webs grossly resemble the cumulative webs in both prior catalogs, but the median $S \times C$ is greater, and the median $\%B$ is lower in the 16 cumulative webs than in either prior catalog. Even for these two statistics, the median value for the 16 cumulative webs falls well within the range of variation of both prior catalogs. The time-specific webs in the 16 cases differ from those of the two prior catalogs somewhat more than do the cumulative webs. Comparisons between time-specific and cumulative versions of a web, one system at a time, are more sensitive than rough comparisons between collections of webs because the methods used to define species and links are (usually) consistent within a study.

6.3 Benthic-pelagic coupling and sediment transport

Benthic-pelagic coupling is a concept in which sea floor processes affect pelagic ecosystems. It involves several oceanographic processes related to the chemistry, biology, and physics that actively link sea floor (benthos) with the overlying water masses (pelagos). The exchange of energy, nutrients, and organisms from one habitat to the other is a coupling of the two independent systems. The deposition of organic matter on the sea floor derived from activities in surface waters such as photosynthesis, sloppy feeding, and excretion is termed benthic-pelagic coupling. The opposite of this action is benthic-pelagic coupling in which materials recycled or created in the sediment returns to the pelagic realm to make a consequential impact on processes and organisms in the surface waters.

Benthic-pelagic coupling is particularly important in shallow coastal and estuarine ecosystems. Reaction rates in sediments tend to be high because organic matter accumulates at the sediment surface, which fuels microbial degradation and mineral cycling. When oxygen depletion occurs within the sediments, alternative compounds are used by microbes as the terminal electron acceptor for respiration. These include NO_3-reduction (denitrification), SO_4^{2-} reduction, and Fe^{3+} reduction, in the order of energetic preference. As a result, anoxic sediments are particularly important zones for N_2O and N_2 production, H_2S production, and also cycling and mobilization of phosphorus, which is linked to the reduction of Fe. A wide variety of models have been developed that can account for both oxic and anoxic sediment biogeochemical reactions. These range from simple parameterizations of the major sedimentary nutrient sources and sinks, to highly detailed reduction-oxidation sediment biogeochemical models.

Nutrient and sediment loading from rivers is also very important in coastal and estuarine ecosystems and therefore must be accounted for in models. These loads can be specified in the boundary conditions as the product of the observed nutrient and/or sediment concentration times the riverine flow, for example $N_{\text{load}} = \text{concentration} \times \text{flow} = \text{mmol NO}_3\,\text{m}^{-3} \times \text{m}^3\,\text{s}^{-1} = \text{mmol NO}_3\,\text{s}^{-1}$. In basin- and global-scale models that do not resolve rivers, inputs of fresh water and nutrients can be specified as an augmentation of the precipitation flux in the vicinity of the outflow. Riverine DOM (Raymond and Spencer, 2015) and seston loading is important because it can have a strong influence on light penetration and therefore primary production in coastal and estuarine waters. They are accounted for in the KCDOM and Kseston terms of the light transmission models. KCDOM can sometimes be approximated using empirical relationships linked to salinity because there is a strong inverse relationship between salinity and CDOM concentrations. Dynamically determining Kseston in shallow estuaries is a significant challenge because it requires implementation of a seston/sediment transport model at some level, that is, with loading as well as deposition on the bottom and resuspension.

Atmospheric nutrient deposition can also be important in both coastal and open ocean ecosystems. For example, in Chesapeake Bay, USA, atmospheric nitrogen deposition comprises ~25% of the total annual N loading. And the atmospheric deposition is the primary source of Fe in the open ocean. These fluxes can be specified from direct measurements or models of atmospheric deposition. However, the former usually requires that the fluxes be specified in a highly idealized way due to sparse observations. Accounting for both wet versus dry deposition can also be problematic if both are not measured.

6.4 Plecoptera (Stoneflies)

In *Encyclopedia of Inland Waters,* DeWalt (2009) investigates ecology of feeding and trophic interactions. Most of what is known about the feeding ecology of stoneflies comes from nymphal studies in the Northern Hemisphere and scattered works elsewhere. Enough is known at this point to say that most stonefly families are detritivorous, feeding on coarse and fine dead plant material and associated biofilms. These films probably constitute the most nutritious component of the diet for detritivores. Confirmed detritivorous families include the Capniidae, Nemouridae, Leuctridae, Notonemouridae, Scopuridae, Taeniopterygidae, Austroperlidae and Diamphipnoidae (wood gougers), Gripopterygidae (some exceptions), Pteronarcyidae, and Peltoperlidae. Predators include the Chloroperlidae, Eustheniidae, Perlidae, Perlodidae, and Styloperlidae. Some Isoperla (Perlodidae) have predaceous mouthparts, but some eat detritus throughout their development. Others experience ontogenetic shifts in diet, moving through detritivory, omnivory, and carnivory. Adult feeding has rarely been studied in detail. Feeding is known for a substantial number of detritivorous species. Leuctridae, Nemouridae, and Capniidae are known to eat cyanolichens, blue-green algae, and the hyphae and spores of Ascomycetes. The Taeniopterygidae appear to be the only family whose adults feed on live, vascular plant tissue, having been implicated in damage to blossoms and leaves of fruit trees. Feeding in the predatory families is apparently limited to the Chloroperlidae and Perlodidae which principally ingest pollens of various types, fine and coarse particulate organic matter, and the spores and hyphae of Ascomycetes. It would not be surprising to find that many more adults in multiple families feed to some extent. Feeding seems to correspond with extended longevity in adults, giving them time to mature eggs and disperse. The use of gut analysis to determine diet has some drawbacks, including difficulty in quantifying items. New methods, such as stable isotope analysis, hold promise to more fully elucidate food webs, especially in detritivorous and omnivorous species, where gut contents are often difficult to categorize beyond miscellaneous detritus.

Niche partitioning: Most studies of stonefly species assemblages have taken place in the Northern Hemisphere, where 30 or more species may emerge from cool water or mountain streams. Genera are composed of multiple species, and densities of each species lead to the potential for space and food competition. Every study of an entire assemblage of stoneflies demonstrates that species partition themselves by serial development and emergence. The researchers depict such a succession of species for a hypothetical assemblage from a moderately large stream in the Midwest, USA. Winter stoneflies emerge from January through March. A small contingent of nemourids, leuctrids, capniids, and perlodid species emerge in April, followed by a brief lull in emergence. By late April, Pteronarcys emerges, as do a perlid and several Isoperla species. In late May, there is another lull before several large perlid species emerge in early June. Small perlids follow in mid-to-late June, July, and early August. The final species to emerge is Leuctra tenuis. No more species emerge until the following January. This sort of succession of species allows for coexistence of species with similar feeding strategies and habitat needs.

A figure is known for hypothetical stonefly community from a moderately large stream in the Midwest, USA. Species list and timing is based on museum specimen records at the Illinois Natural History Survey.

Stoneflies as indicators of water quality: Stoneflies are highly sensitive to organic pollution and hypoxia that comes with it. Biotic indices have been developed throughout the world that consistently rate stoneflies as most intolerant to pollution. Although several of these ratings systems have been developed through professional judgment, others have empirically generated tolerance values. Even these have rated stoneflies as the most sensitive order of aquatic insects. Throughout the world, the number of species of Ephemeroptera (mayflies), Plecoptera, and Trichoptera (caddisflies) (EPT taxa) is used as a stream quality metric. Unfortunately, many water pollution biologists these days are increasingly finding only mayflies or caddisflies totally.

Conservation ecology: Freshwater aquatic systems have suffered 4–5 × higher extinction rates than terrestrial habitats, and this is likely to continue into the future. In Europe, large river species have lost range or have been extirpated from many countries. There is even fear that European mountain habitats that once provided refugia for species, and promoted gene flow due to interconnectivity, are becoming fragmented due to acid rainfall and other disturbances. The Nature Conservancy and its spin-off, Nature Serve, have suggested that in the USA, stoneflies are the third most imperiled grouping of organisms both in terms of proportion of species imperiled (behind mussels and crayfish) and in the total number of imperiled species (behind vascular plants and mussels). In the highly agricultural, glaciated landscapes in Illinois, USA, stoneflies have a higher extirpation rate than for either mussels or fishes. It is clear that human needs for water, conversion of land to farming, industry, and housing, and climate change put stoneflies and a lot of other aquatic fauna at risk. In many areas, diapausing species are replacing the once widespread and common univoltine-slow and semivoltine species. Families that are particularly hard hit in the North American fauna are in the Perlidae and Perlodidae, and in the Midwest, these were lost during a relatively short time in the late 1940s through the early 1960s. This is the same time period as for egg shell thinning in raptors, suggesting that indiscriminant DDT usage may be at fault. Evidence is currently unfolding that suggests that global climate change is having an effect upon stonefly distributions. In the Great Smoky Mountains National Park, USA, it appears that at least one large perlid stonefly, Acroneuria abnormis, has significantly increased its upper altitudinal construction. However, several other species have narrower altitudinal distributions in the region and if both the lower and higher boundaries are shifted upward, these species may run into a ceiling beyond which they cannot colonize. Limits include the height of the local mountains or the altitude at which acidification of streams becomes problematic, something that is already occurring in the mountains of the eastern USA and in Europe.

Food-chain length is an important character of ecological communities that affects many of their functional aspects. Recently, an increasing number of studies have tested the effects of productivity, disturbance, or ecosystem size on food-chain length in a variety of natural systems. Here, we conduct a formal metaanalysis to summarize findings from these empirical studies. We found significant positive mean effects of productivity and ecosystem size but no significant mean effect of disturbance on food-chain length. The strength of mean effect sizes was not significantly different between productivity and ecosystem size. These results lend general support to previous theories predicting the effect of productivity and ecosystem size, but fail to support the prediction that disturbance shortens food chains. In addition, a metaanalysis found that the effect sizes of primary studies were significantly heterogeneous for ecosystem size and disturbance, but not for productivity. This pattern might reflect that ecosystem size and disturbance can affect food-chain length through multiple different mechanisms, while productivity influences food-chain length in a simple manner through energy limitation. Food has been one of the major drivers of human interaction with coastal and estuarine ecosystems, and this has impelled much of the study of these systems. In its initial phase, greater understanding would have brought direct payoffs in the form of greater catches and, now, greater understanding is necessary to not just to counter past mistakes (such as overexploitation, habitat loss, and pollution) but also future pressures (such as climate change). The key change has been in the recognition that the oceans are not an infinite resource and that sustainability of the resources is the key not only regarding commercial returns, but also socioeconomic stability of the human populations. As a result, scientific investigations into the cause of exploitation of the fisheries' resources, for example, have progressed from a simple goal of maximizing the catches through one where long-term yields (e.g., MSY or maximum sustainable yield) of single species were the required model outputs to the present situation in which the sustainability of the ecosystem itself is the overriding goal.

6.5 Mangrove trophic interactions and estuarine ecosystems

Research on mangrove trophic interactions has rarely been framed as a test of theory in food-web ecology and, as a consequence, is seldom cited in textbooks or synthetic reviews as illustrating general principles. It is our hope that an informative publication will help bring past and emerging research on mangrove trophic interactions to the attention of ecologists working in other habitats. As a study system, mangroves hold great promise for the general ecological theory. For example, the environmental setting is tailor-made for investigations of the importance of trophic subsidies among habitats to local population and community dynamics, that is, reciprocal exchanges among mangrove, coastal marine, riverine, and terrestrial habitats. Experimental manipulations of resource availability and the densities of consumers have demonstrated both bottom-up resource limitation and strong top-down consumer control in different compartments of the food web. Crabs are particularly strong interactors, exerting top-down control over litter dynamics and seedling recruitment in numerous mangrove forests. Finally, the confirmation that mangroves are key sites of carbon storage has major implications for the prioritization of habitat protection and management efforts in response to accelerating climate change. habitats to local population and community dynamics (that is, reciprocal exchanges among mangrove, coastal marine, riverine, and terrestrial habitats). Read Heymans et al. (2011) chapter 9.06 on ecopeath theory, modeling, and application to coastal ecosystems, in Vol. 9 of Treatise on Estuarine and Coastal Science. Study the problem of trophic interactions in coastal and estuarine mangrove forest ecosystems as reported by Sousa and Dangremond (2011) in Vol. 6 of Trophic Relationships of Coastal and Estuarine Ecosystems.

Ecosystem models describe trophic interactions within the ecosystems and provide a good basis for studying the general patterns of ecological properties. We can read a review of 75 ecopath models of coastal ecosystems to describe and assess their structural and functional characteristics and to investigate the ecological roles of their main functional groups. The analysis highlights the influence of depth, latitude, and longitude on their main properties, the importance of different ecosystem types in distinguishing different ecological features, and the influence of the total size of the modeled ecosystem on its properties.

Fig. 6.5 describes relation between ecological system and social system. These data were provided by Livingston (2000) and Heymans et al. (2011).

6.6 Spatial aspects of food webs

Brose et al. (2005a, b) write in *Dynamic Food Webs*: "Trophic interactions are key determinants of population abundance and dynamics, the structure and persistence of communities, and the rate and sustainability of ecosystem processes," see also (Ings et al., 2009). Ecological models show that complexity usually destabilizes food webs (May, 1971, 1973) predicting that food webs should not amass the large numbers of interacting species that are found in nature (Winemiller, 1990; Polis, 1991; Lavigne, 1995). In particular, Winemiller (1990) analyze spatial and temporal variation in tropical fish trophic networks. Using nonlinear models, one may study the influence of interaction strength (likelihood of consumption of one species by another) on food-web dynamics away from equilibrium. Consistent with previous suggestions (Gardner and Ashby, 1970; May, 1971, 1973), the more recent results (Hastings and Powell, 1991; McCann et al., 1998) show that weak-to-intermediate strength links are important in promoting community persistence and stability. Weak links act to dampen oscillations between consumers and resources. This tends to maintain population densities further away from zero and to decrease the statistical chances that a population will become extinct (lower population densities are more prone to such chances). Data on interaction strengths in natural food webs (Fagan and Hurd, 1994; Wootton, 1997; Goldwasser and Roughgarden, 1993; Paine, 1992; Rafaeli and Hall, 1996) indicate that food-web interaction strengths are indeed characterized by many weak interactions and a few strong interactions.

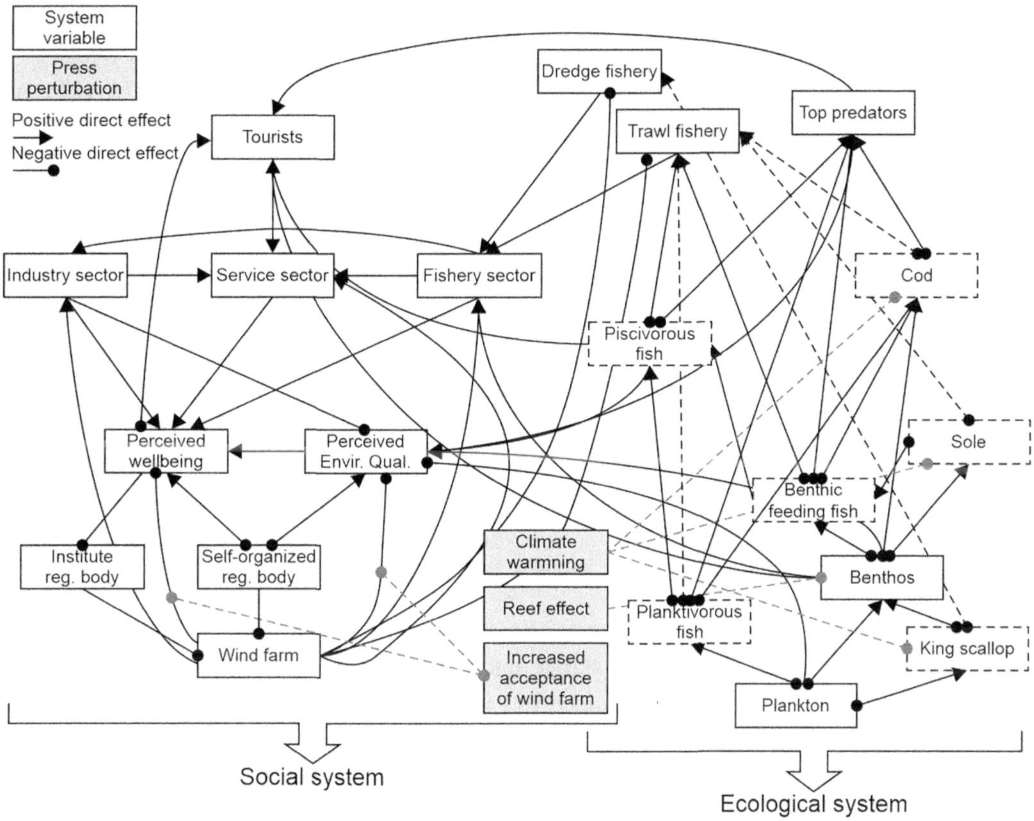

Fig. 6.5

Relation between ecological system and social system. A visualization of difference between complex computer simulations and statistical models. Dynamic biodiversity maintenance in food webs (oyster and mussel beds) is treated by Livingston (2000), de Ruiter and Moore (2004), Heymans et al., 2011, and some others.

6.7 Summary

Summing up, we can ensure that the study of food webs has been a central theme within ecology for decades, and their structure and dynamics have been used to assess a range of key properties of communities (e.g., complexity-stability relationships) and ecosystems (e.g., fluxes of energy and nutrients). However, many food web parameters are sensitive to sampling effort, which is seldom considered, and, most studies have used either species- or size-averaged data for both nodes and links, rather than individual-based data, which is the level of organization at which trophic interactions occur. This practice of aggregating data hides a considerable amount of biologically meaningful variation and could, together with potential sampling effects, create methodological artifacts. New individual-based approaches could improve the understanding of, and ability to predict, food-web structure and dynamics, particularly if they are derived from simple metabolic and foraging constraints. Dedicated researchers explored the effect of species-averaging in four highly-resolved individual-based aquatic food webs (Broadstone Stream, the Afon Hirnant, Tadnoll Brook and the Celtic Sea) and found that it obscured structural regularities resulting from intraspecific size variation. The individual-based approach provided clearer insights into seasonal and ontogenetic shifts, highlighting the importance of the temporal component of size-structuring in ecological networks. An extension of the Allometric Diet Breadth Model predicted the structure of empirical food webs almost twice as accurately as the equivalent species-based webs, with the best-fitting model predicting 83% of the links correctly in the Broadstone Stream size-based web, and the few mismatches between the model and data can be explained largely by sampling effects. The results highlight the need for theoretical explanations to correspond closely with methods of data collection and aggregation. Suggestions appeared how this situation can be improved by including individual-level data and more explicit information on sampling effort when constructing food webs in future studies. Allhoff et al. (2015) developed evolutionary food-web model based on body masses, which gives realistic networks with permanent species turnover.

Food webs, the networks of predator-prey interactions in ecological systems, are remarkably complex, but nevertheless surprisingly stable in terms of long-term persistence of the system as a whole. See Rosenzweig and McArthur (1963) for their graphical representation and stability conditions of predator-prey interactions. In order to understand the mechanism driving the complexity and stability of food webs, eco-evolutionary food-web modeling has been developed in which new species emerge via evolution from existing ones and dynamic ecological interactions determine which species is viable and which goes extinct. The food-web structure thereby emerges from the dynamical interplay between speciation and trophic interactions. The model is less abstract than earlier models of this type in the sense that all three evolving traits have a clear biological meaning, namely the average body size of the individuals, the preferred prey body mass, and the width of their potential prey body mass spectrum. Networks may be observed with a wide range of sizes and structures that show a high similarity to natural food webs. These networks exhibit a continuous species turnover, during which phases of increasing specialization are interrupted by extinction avalanches. After these avalanches, more generalized predators can again occur. Massive extinction waves that could affect the network are not observed, suggesting that related events in earth's history had external causes (Brose et al., 2005a, b). Until recently, aspects of spatial scale have been largely ignored in empirical and theoretical food web studies (e.g., Berg and Bengtsson, 2007; Cohen and Briand, 1984; Martinez, 1992). Most ecologists tend to conceptualize and represent food webs as static representations of communities, depicting a community assemblage as sampled at a particular point in time, or highly aggregated trophic group composites over broader scales of time and space (Polis and Strong, 1996). Moreover, most researchers depict potential food webs, which contain all species sampled and all potential trophic links based on literature reviews, several sampling events, or laboratory feeding trials. In reality, however, not all these potential feeding links are realized as not all species co-occur. Also, not all samples in space or time can contain all species (Schoenly and Cohen, 1991), hence a variance of food web architecture in space (Brose et al., 2005a, b).

In recent years, food-web ecologists have recognized that food webs are open systems influenced by processes in adjacent systems that may be spatially heterogeneous (Polis and Strong, 1996). This influence can be bottom-up, due to allochthonous inputs of resources. Polis (1991) also considered complex trophic interactions in deserts in the context of an empirical critique of the food-web theory. Lavigne (1995) summarized effects of interactions between marine mammals and their prey in the context of thermal aspects accompanying these interactions.

Polis and Strong (1996) tested diverse food webs achieving the following conclusions: Food webs in nature have multiple, reticulate connections between a diversity of consumers and resources. Such complexity affect web dynamics: it first spreads the direct effects of consumption and productivity throughout the web rather than focusing them at the particular "trophic levels." Second, consumer densities are often donor controlled with food from across the trophic spectrum, the herbivore and detrital channels, other habitats, life-history omnivory, and even trophic mutualism. Although consumers usually do not affect these resources, increased numbers often allow consumers to depress other resources to levels lower than if donor-controlled resources were absent. It has been proposed that such donor-controlled and "multichannel" omnivory is a general feature of consumer control and central to food-web dynamics. This observation is contrary to the normal practice of inferring dynamics by simplifying webs into few linear "trophic levels" as per "green world" theories. An opinion appeared that such theories do not accommodate common and dynamically important features of real webs such as the ubiquity of donor control and the importance and dynamics of detritus, omnivory, resources crossing habitats, life history, nutrients (as opposed to energy), pathogens, resources defenses, and trophic symbioses. The conclusion was that trophic cascades and top-down community regulation as envisioned by trophic-level theories are relatively uncommon in nature.

Polis and Winemiller (2013) consider food webs as tools for the integration of patterns and dynamics in the context of ecology and evolution. Their 2013 book generalizes in a sense an earlier Winemiller's book on the spatial and temporal variation in tropical-fish trophic networks (Winemiller, 1990). See also: Huxel and McCann (1998), Mulder and De Zwart (2003), or any top-down due to the regular or irregular presence of top predators (e.g., Post and Takimoto, 2007; Post et al., 2000). However, without a clear understanding of the size of a system and a definition of its boundaries, it is not possible to judge whether flows are internal or driven by adjacent systems. Similarly, the importance of allochthony is only assessable when the balance of inputs and outputs are known relative to the scale and throughputs within the system itself. At the largest scale of the food web—the home range of a predator such as wolf, lion, shark or eagle of roughly 50–300 km^2—the balance of inputs and outputs caused by wind and movement of water may be small compared with the total trophic flows within the home range of the large predator (Cousins, 1990). Acknowledging these issues of space, Polis and Strong (1996) argue that a progress toward the next phase of food-web studies would require addressing spatial and temporal processes. Here is a conceptual framework with some nuclei about the role of space in food-web ecology. Although primarily spatial aspects are addressed, this framework is linked to a more general concept of spatiotemporal scales of ecological research (Hastings, 2005).

When formulating proximate structural mechanisms for variation in food-chain length, Post and Takimoto (2007) as well as Takimoto and Post (2012) regard the food-chain length as a central characteristic of ecological communities because of its strong influence on community structure and ecosystem function. While recent studies have started to better clarify the relationship between food-chain length and environmental gradients such as resource availability and ecosystem size, much less progress has been made in isolating the ultimate and proximate mechanisms that determine food-chain length. Progress has been slow, in part, because the research has paid little attention to the proximate changes in food-web structure that must link variation in food-chain length to the ultimate dynamic mechanism. Taking into account all the above, Post and Takimoto (2007) as well as Takimoto and Post (2012) proposed structural mechanisms that determine variation in food-chain length.

One could explore the implications of these mechanisms for understanding how changes in food-web structure influence food-chain length when using both an intraguild predation community model and data from natural ecosystems. The resulting framework provides the mechanisms for linking ultimate dynamic mechanisms to variation in food-chain length. It also suggests that simple linear food-chain models may make misleading predictions about patterns of variation in food-chain length because they are unable to incorporate important structural mechanisms that alter food-web dynamics and cause nonlinear shifts in the food-web structure. Intraguild predation models provide a more appropriate theoretical framework for understanding food-chain length in most natural ecosystems because they accommodate all of the proximate structural mechanisms, as identified by Post and Takimoto (2007) and Takimoto and Post (2012, 2013).

As stated by Dunne et al. (2004), previous studies suggest that food-web theory has yet to account for major differences in food-web properties of marine versus other types of ecosystems. Dunne et al. (2004) examined this issue by analyzing the network structure of food webs for the Northeast US Shelf, a Caribbean reef, and Benguela, off South Africa. The values of connectance (links per species), link density (links per species), mean chain length, and fractions of intermediate, omnivorous, and cannibalistic taxa of these marine webs are somewhat high but still within the ranges observed in other webs. Dunne et al. (2004) further compared the marine webs by using the empirically corroborated niche model that accounts for observed variation in diversity (taxon number) and complexity (connectance). Their results substantiate previously reported results for estuarine, freshwater, and terrestrial data sets, which suggests that food webs from different types of ecosystems with variable diversity and complexity share fundamental structural and ordering characteristics. Analyses of potential secondary extinctions resulting from species loss show that the structural robustness of marine food webs is also consistent with trends from other food webs. As expected, given their relatively high connectance, marine food webs appear fairly robust to loss of most-connected taxa as well as random taxa. Still, the short average path length between marine taxa (1.6 links) suggests that effects from perturbations, such as overfishing, can be transmitted more widely throughout marine ecosystems than previously observed (Fig. 6.6).

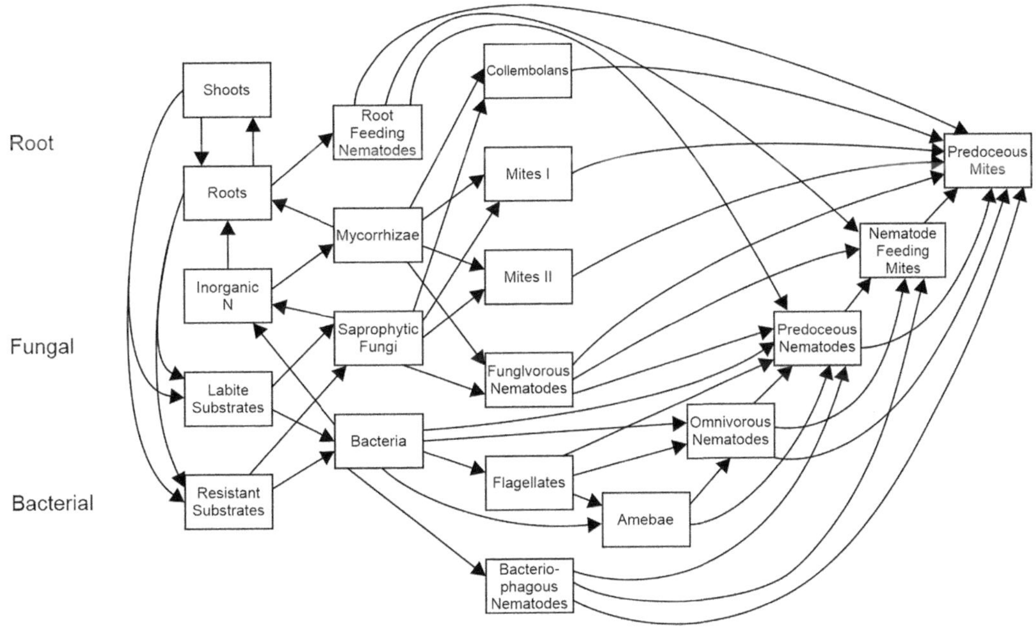

Fig. 6.6

Relation between coupling with a community in solid food webs depicting the fungal, root, and bacterial compartments, with the tendency for higher order consumers coupling these pathways. Here the fungal and bacterial channels tend to derive from recalcitrant and labile pathways, respectively.

References

Allhoff, K.T., Ritterskamp, D., Rall, B.C., Drossel, B., Guill, C., 2015. Evolutionary food web model based on body masses gives realistic networks with permanent species turnover. Scientific Rep. 5 (1). https://doi.org/10.1038/srep10955. ArXiv. Article number: 10955.

Asmus, H., Asmus, R.M., Prins, T.C., 1992. Benthic-pelagic flux rates on mussel beds: tunnel and tidal flume methodology compared. Helgolander Meeresunters 46, 341–361. https://doi.org/10.1007/BF02367104.

Asmus, H., Asmus, R., 2000. Material exchange and food web of seagrass beds in the Sylt-Rømø Bight: how significant are community changes at the ecosystem level? Helgol. Mar. Res. 54, 137–150. https://doi.org/10.1007/s101520050012.

Asmus, H., Asmus, R., 2005. Significance of suspension-feeder systems on different spatial scales. In: The Comparative Roles of Suspension-Feeders in Ecosystems, pp. 199–219, https://doi.org/10.1007/1-4020-3030-4-12.

Asmus, H., Asmus, R., 2011. General aspects of suspension-feeder communities. In: McLusky, D., Wolanski, E. (Eds.), Treatise on Estuarine and Coastal Science. Academic Press—Elsevier. ISBN-10: 0123747112, ISBN-13: 978-0123747112.

Bagdassarian, C.K., Dunham, A.E., Brown, C.G., Rausher, D., 2007. Biodiversity maitenance in food webs with regulatory environmental feedback. J. Theor. Biol. 245, 705–714.

Baird, M.E., Suthers, I.M., 2007. A size-resolved pelagic ecosystem model. Ecol. Model. 203 (3), 185–203. https://doi.org/10.1016/j.ecolmodel.2006.11.025.

Baird, M.E., Asmus, H., Asmus, R., 2007. Trophic dynamics of eight intertidal communities of the Sylt-Rømø Bight ecosystem, northern Wadden Sea. Mar. Ecol Prog. Ser. 351, 25–41.

Berg, M.R., Bengtsson, J., 2007. Temporal and spatial variability in solid food web structure. OIKOS 116 (11), 1799–1804.

Brose, U., Berlow, E.L., Martinez, N.D., 2005a. Scaling up keystone effects from simple to complex ecological networks. Ecol. Lett. 8, 1317–1325. https://doi.org/10.1111/j.1461-0248.2005.00838.x.

Brose, U., Pavao-Zuckerman, M., Eklöf, A., Bengtsson, J., Berg, M., Cousins, S.H., Mulder, C., Verhoef, H.A., Wolters, V., 2005b. In: de Ruiter, P.C., Wolters, V., Moore, J.C. (Eds.), Dynamic Food Webs: Multispecies Assemblages, Ecosystem Development and Environmental Change. vol. 3. Elsevier, London, UK, pp. 463–469.

Cohen, E., Briand, F., 1984. Trophic links of community food webs. Proc. Natl Acad. Sci. USA, 81, 4105–4150.

Commito, J.A., Boncavage, E.M., 1989. Suspension-feeders and coexisting infauna: an enhancement counterexample. J. Exp. Mar. Biol. Ecol. 125, 33.

Commito, J.A., Celano, E.A., Celico, H.J., Como, S., Johnson, C.P., 2008. Species diversity in the soft-bottom intertidal zone: biogenic structure, sediment, and macrofauna across mussel bed spatial scales. J. Exp. Mar. Biol. Ecol. 366, 70–81.

Cousins, D.H., 1990. Plankton production and year—class strength in fish populations: an update of the match/mismatch hypothesis. Adv. Mar. Biol. 26, 249–293.

Dame, R., Dankers, N., Prins, T., Jongsma, H., Smaal, A., 1991a. The influence of mussel beds on nutrients in the Western Wadden Sea and Eastern Scheldt Estuaries. Estuaries 14 (2), 130–138.

Dame, R.F., Spurrier, J.D., Williams, T.M., Kjerfve, B., Zingmark, R.G., Wolaver, T.G., Chrzanowski, T.H., McKellar, H.N., Vernberg, F.J., 1991b. Annual material processing by a salt Marsh-estuarine basin in South Carolina, USA. Mar. Ecol. Progr. Ser. 72 (1/2), 153–166. http://www.jstor.org/stable/24825438.

DeWalt, 2009. In: Trockner, K., Mechner, T. (Eds.), Encyclopedia of Inland Waters, second ed. Elsevier, Amsterdam.

Dunne, J.A., Williams, R.J., Martinez, N.D., 2004. Network structure and robustness of marine food webs. Mar. Ecol. Prog. Ser. 273, 291–302 (pp. 3661–3682 in 'Food Webs').

Elton, C., 1927. Anmal Ecology. Sidgwick & Jackson, London.

Fagan, W.F., Hurd, L.E., 1994. Hatch density variation of a generalist athropod predator: population consequences and community impact. Ecology 75, 2022–2032.

Gardner, M., Ashby, W.R., 1970. Connectance of large dynamical (cybernetic) systems: critical value for stability. Nature 228, 784.

Goldwasser, L., Roughgarden, J., 1993. Construction and analysis of a large caribbean food web. Ecology 74, 1216–1233.

Hastings, A. (Ed.), 2005. Theoretical Ecology Series. vol. 3. Elsevier, London, UK, pp. 463–469. ISSN 1875-306X URN: urn:nbn:se:liu:diva-31067 Local ID: 16789ISBN: 9780120884582 (print), ISBN: 0120884585.

Hastings, A., Powell, T., 1991. Chaos in a three species food chain model. Ecology 72, 896–903.

Heymans, J.J., Coll, M., Libralato, S., Christensen, V., 2011. In 9.06—ecopath theory, modeling, and application to coastal ecosystems. Treatise Estuar. Coast. Sci. 9, 93–113. https://doi.org/10.1016/B978-0-12-374711-2.00905-0.

Hillel, D. (Ed.), 2004. Encyclopedia of Soils in the Environment. Academic Press. Imprint: ISBN: 9780123485304.

Huxel, G., McCann, K., 1998. Food web stability: the influence of trophic flows across habitats. Am. Nat. 152, 460–469.

Ings, T.C., Montoya, J.M., Bascompte, J., Blüthgen, N., Brown, L., Dormann, C.F., Edwards, F., Figueroa, D., Jacob, U., Iwan Jones, J., Lauridsen, R.B., Ledger, M.E., Lewis, H.M., Olesen, J.M., Frank Van Veen, F.J., Warren, P.H., Woodward, G., 2009. Review: ecological networks—beyond food webs. J. Anim. Ecol. 78 (1), 253–269.

Lavigne, D.M., 1995. Interactions between marine mammals and their prey: unravelling the tangled web. In: Montevecchi, W.A. (Ed.), Studies of High Latitude Homeotherms in Cold Ocean Systems. Canadian Wildlife Service Occasional Paper, Ottawa, Canada, pp. 95–102.

Lindeman, R.L., 1942. The trophic-dynamic aspects of ecology-effect of environmental feedback. Ecology 23 (4), 399–417.

Livingston, R.J., 2000. Eutrophication Processes in Coastal Systems. CRC Press, ISBN: 9780849390623. https://www.routledge.com/Eutrophication-Processes.

Martinez, N.D., 1992. Constant connectance in community food webs. Am. Nat. 139 (6), 1208–1218.

May, R.M., 1971. Stability in multispecies community models. Math. Biosci. 12, 59–79.

May, R.M., 1973. Stanility and Complexity in Model Ecosystems. Princeton University Press, Princeton,NJ.

McCann, K., Hastings, A., Huxel, G.R., 1998. Weak trophic interactions and the balance of nature. Nature 395 (6704), 794–798. https://doi.org/10.1038/27427.

Mulder, Ch, De Zwart, D., 2003. Assesing fungal species sensivity to environmental gradients by the Elllenberg indicator values of above-ground vegetation. Basic Appl. Ecol. 4 (6), 557–568.

Paine, R.T., 1992. Food-web analysis through field measurement of per capita interaction strength. Nature 355, 73–75.

Polis, G.A., 1991. Complex trophic interactions in deserts: an empirical critique of food web theory. Am. Nat. 138, 123–155.

Polis, G.A., Strong, D., 1996. Food web complexity and community dynamics. Am. Nat. 147, 813–846.

Polis, G.A., Winemiller, K.O., 2013. Food Webs: Integration of Patterns & Dynamics. Springer Science & Business Media, https://doi.org/10.1007/878-4615-7007-3. ISBN: 1461570077, 9781461570073.

Post, D.M., Takimoto, G., 2007. Proximate structural variation in food-chain length. OIKOS 116 (5), 775–782.

Post, D.M., Pace, M.L., Hairston, N.G., 2000. Ecosystem size determines food-chain length in lakes. Nature 405, 1047–1049.

Prins, T.C., Smaal, A.C., 1990a. Benthic-pelagic coupling: the release of inorganic nutrients by an intertidal bed of *Mytilus edulis*. In: Bames, M., Gibson, R.N. (Eds.), Trophic Relationships in the Marine Environment, Proc 24 Europ. Mar. Biol. Symp. Aberdeen University Press, pp. 89–103.

Prins, T.C., Smaal, A.C., Dame, R.F., 1998. A review of the feedbacks between bivalve grazing and ecosystem processes. Mar. Ecol. Prog. Ser. 142, 121–134.

Prins, T.C., Smaal, A.C., 1990b. Benthic-pelagic coupling: the release of inorganic nutrients by an intertidal bed of *Mytilus edulis*. In: Barnes, M., Gibson, R.N. (Eds.), Trophic Relationship in the Marine Environment. Proc. 24th Europ. Mar. Biol. Symp., Aberdeen University Press, pp. 89–103.

Rafaeli, D.G., Hall, S.J., 1996. In: Polis, G.A., Winemiller, K.O. (Eds.), Food Webs: Integration of Patterns & Dynamics. Chapman & Hall, New York, pp. 185–191.

Raymond, P.A., Spencer, R.G.M., 2015. In: Hansell, D.A., Carlson, C.A. (Eds.), 'Riverine DOM' Biogeochemistry of Marine Dissolved Organic Matter, second ed. Elsevier, Amsterdam, pp. 509–533, https://doi.org/10.1016/b978-0-12-405940-5.00011-x. See also: Sci. Tech. Connect—Elsevier, http://scitechconnect.elsevier.com, p. 509.

Rosenzweig, M., McArthur, R.H., 1963. Graphical representation and stability conditions of predator-prey interactions. Am. Nat. 107, 275–294.

de Ruiter, P.C., Moore, J.C., 2004. Trophic interactions and the dynamics and stability of soil populations and food webs. In: Encyclopedia of Soils in the Environment. Academic Press, ISBN: 9780123485304.

Schoenly, K., Cohen, J.E., 1991. Temporal variation in food web structure: 16 empirical cases. Ecol. Monogr. 61 (3), 267–298. https://doi.org/10.2307/2937109.

Smaal, A.C., Haas, H.A., 1997. Seston dynamics and food availability on mussel and cockle beds. Estuar. Coast. Shelf Sci. 45 (2), 247–259. https://doi.org/10.1006/ecss.1996.0176.

Smaal, A.C., Zurburg, W., 1997. The uptake and release of suspended and dissolved material by oysters and mussels in Marennes-Oléron Bay. Aquat. Living Resour. 10 (1). https://doi.org/10.1051/alr:1997003.

Sousa, W.P., Dangremond, E.M., 2011. In: Wolanski, E. (Ed.), Trophic Interactions in Coastal and Estuarine Mangrove Forest Ecosystems. Vol. 6 of Trophic Relationships of Coastal and Estuarine Ecosystems, Elsevier-Academic Press, p. 43. Wilson, J.G., Luczkovich, J.J. (Vol. Eds.), Luczkovich Treatise on Estuarine and Coastal Science.

Takimoto, G., Post, D.M., 2012. Environmental determinants of food-chain length: a meta-analysis. Ecol. Res. 28 (5). https://doi.org/10.1007/s11284-012-0943-7.

Takimoto, G., Post, D.M., 2013. Environmental determinants of food-chain length: a meta-analysis. Ecol. Res. 28, 675–681. https://doi.org/10.1007/s11284-012-0943-7.

Winemiller, K.O., 1990. Spatial and temporal variation in tropical fish trophic networks. Ecol. Monogr. 60, 331–367.

Wootton, J.T., 1997. Estimates and test of per capita interaction strength: diet, abundance, and impact of intertidally foraging birds. Ecol. Monogr. 67, 45–64.

Further reading

Asmus, H., Asmus, R.M., Reise, K., 1990. Exchange processes in an intertidal mussel bed: a Syltflume study in the Wadden Sea. Bet. Biol. Anst. Helgoland 6, 1–79.

Chesson, J., 1983. The estimation and analysis of preference and its relationship to foraging models. Ecology 64, 1297–1304.

De Feo, O., Rinaldi, S., 1998. Singular homoclinic bifurcations in tritrophic food chains. Math. Biosci. 148, 7–20.

Diehl, S., 1993. Relative consumer sizes and the strengths of direct and indirect interactions in omnivorous feeding relationships. Oikos 68, 151–157.

Elton, C.S., 1958. Ecology of Invasions by Animals and Plants. Chapman & Hall, London.

Fagan, W.F., 1997. Omnivory as a stabilizing feature of natural communities. Am. Nat. 150, 554–567.

Holt, R.D., 1977. Predation, apparent competition, and the structure of prey communities. Theor. Popul. Biol. 12, 197–229.

Holt, R.D., 1996. In: Begon, M., Gange, A., Brown, V. (Eds.), In Multitrophic Interactions. Chapman & Hall, London, pp. 333–350.

MacArthur, R.H., 1955. Fluctuations of animal populations and a measure of community stability. Ecology 36, 533–536.

McCann, K., Hastings, A., 1997. Re-evaluating the omnivory-stability relationship in food webs. Proc. R. Soc. Lond. B 264, 1249–1254.

McCann, K., Yodzis, P., 1995. Biological conditions for chaos in a three-species food chain. Ecology 75, 561–564.

Polis, G.A., 1994. Food webs, trophic cascades and community structure. Aust. J. Ecol. 19, 121–136.

Polis, G.A., Holt, R.D., 1992. Intraguild predation: the dynamics of complex trophic interactions. Tree 7, 151–155.

Prins, T.C., Escaravage, V., Pouwer, A.J., Wetsteyn, L.P.M.J., Haas, H.A., 1997a. Description of mesocosms and methods, and a comparison with North Sea conditions. In: Smaal, A.C., Peeters, J.C.H., Prins, T.C., Haas, H.A., Heip, C.H.R. (Eds.), The Impact of Marine Eutrophication on Phytoplankton, Zooplankton and Benthic Suspension Feeders. National Institute for Coastal and Marine Management, Middelburg, pp. 19–40. Report RIKZ97.035, NIOO/CEMO-1997.05.

Simenstad, C., Yanagi, T., 2011b. In: Wolanski, E., McLusky, D. (Eds.), Treatise on Estuarine and Coastal Science Classification of Estuarine and Nearshore Coastal Ecosystems.

Simenstad, C., Yanagi, T., 2011. Introduction to classification of estuarine and nearshore coastal ecosystems. Environ. Sci. https://doi.org/10.1016/B978-0-12-374711-2.00101-7.

Strong, D., 1992. Are trophic cascades all wet? Differentiation and donor-control in speciose ecosystems. Ecology 73, 747–754.

Yodzis, P., Innes, S., 1992. Body-size and consumer-resource dynamics. Am. Nat. 139, 1151–1175.

Asmus, H., Asmus, R.M., 1990. Trophic relationships in tidal flat areas: to what extent are tidal flats dependent on imported food? Neth. J. Sea Res. 27, 93–99.
Asmus, R.M., Asmus, H., 1991. Mussel beds: limiting or promoting phytoplankton? J. Exp. Mar. Biol. Ecol. 148, 215–232.
Asmus, H., Asmus, R.M., 1993. Phytoplankton—mussel bed interactions in intertidal ecosystems. In: Dame, R.F. (Ed.), Bivalve Filter-feeders in Estuarine and Coastal Ecosystem Processes. Springer, New York, USA, pp. 57–84.
Asmus, R.M., Jensen, M.H., Jensen, K.M., Kristensen, E., Asmus, H., Wille, A., 1998. The role of water movement and spatial scaling for measurement of dissolved inorganic nitrogen fluxes in intertidal sediments. Estuar. Coast. Shelf Sci. 46, 221–223.
Dame, R.F., Prins, T.C., 1997. Bivalve carrying capacity in coastal ecosystems. Aquat. Ecol. 31, 409–421. https://doi.org/10.1023/A:100999701158.
Prins, T.C., Smaal, A.C., 1994a. The role of the blue mussel *Mytilus edulis* in the cycling of nutrients in the Oosterschelde estuary (The Netherlands). Hydrobiologia 283, 413–429.
Prins, T., Escaravage, V., 2005. Can bivalve suspension-feeders affect pelagic food web structure? In: Dame, R.F., Olenin, S. (Eds.), The Comparative Roles of Suspension-Feeders in Ecosystems. Springer, Dordrecht, pp. 31–51.
Prins, T.C., Smaal, A.C., Dame, R.F., 1997. A review of the feedbacks between bivalve grazing and ecosystem processes. Aquat. Ecol. 31, 349–359.
Prins, T.C., Smaal, A.C., 1994b. The role of the blue mussel *Mytilus edulis* in the cycling of nutrients in the Oosterschelde estuary (The Netherlands). Hydrobiologia 282 (283), 413–429.
Wilson, J.G., Luczkovich, J.J. (Eds.), 2011. Treatise on Estuarine and Coastal Science. Elsevier-Academic Press. Vol. 6 on Trophic Relationships of Coastal and Estuarine Ecosystems. In 6.01: Introduction to Food Webs in Coastal and Estuarine Ecosystems.

CHAPTER 7

Dynamic food webs

7.1 Introduction

7.1.1 Multispecies assemblages, ecosystem development, and environmental change

When describing in the previous chapter coastal and nearshore estuarine ecosystems, we have cited, among others, the publication of de Ruiter and Moore (2004), discussing in the context of food webs the link between the trophic interactions and the stability of soil populations. In the present chapter, we shall characterize a sister publication, in fact, a 2005 book authored by de Ruiter, Wolters, and Moore, briefly cited here as de Ruiter et al. (2005), which treats the dynamic food webs. This is a voluminous treatise, which involves multispecies, assemblages, and ecosystem development subject to environmental changes.

Food webs are special descriptions of biological communities focusing on trophic interactions between consumers and resources. Food webs have become a central tools in population, community, and ecosystem ecology. The interactions within food webs are thought to influence the dynamics and persistence of many populations in fundamental ways through the availability of resources (i.e., energy/nutrients) and the mortality due to predation. Moreover, food web structure and ecosystem processes, such as the cycling of energy and nutrients, are deeply interrelated in that the trophic interactions represent transfer rates of energy and matter. Food webs therefore provide a way to analyze the interrelationships between community dynamics and stability and ecosystem functioning and how these are influenced by environmental change and disturbance (de Ruiter et al., 2005).

Scientists long ago observed how the distribution, abundance, and behavior of organisms are influenced by interactions with other species. Population dynamics of interacting predators and prey are difficult to predict, and many ecosystems are known to contain hundreds or thousands of these interactions arranged in highly complex networks of direct and indirect interactions. Motivated in part by May's (1972) theoretical study of the complexity-stability relationship, the study of food webs gained momentum in the late 1970s and early 1980s (Cohen, 1978; Pimm, 1982). A formal means of dealing with the flow of energy and matter in food webs was ushered with the advent of ecosystem ecology (Odum, 1963), and since then, the food-web approach has been adopted to analyze interrelationships between community structure, stability, and ecosystem processes (DeAngelis, 1992).

Complexity and Complex Ecological Systems. https://doi.org/10.1016/B978-0-443-19237-1.00004-1

As reminded by de Ruiter et al. (2005), the first food-web symposium was convened at Gatlinburg, North Carolina, in 1982 (DeAngelis et al., 1982). That symposium was dominated by theoretical studies focused on the complexity-stability relationship and empirical studies examining features of simple topological webs (ball and stick diagrams) compiled from the published literature. The ensuing decade was marked by exploration of a greater number of issues influencing the structure and dynamics of food webs (interaction strength, indirect effects, keystone species, spatial variation, and temporal variation in abiotic drivers) and a search for more detailed and accurate food-web descriptions (de Ruiter et al., 2005). Some ecologists questioned the utility of analyzing features of web diagrams that quite obviously contained too few taxa, grossly unequal levels of species aggregation, and feeding links with no magnitudes or spatiotemporal variation (Hall and Raffaelli, 1997).

A second food-web symposium, held at Pingree Park, Colorado, in 1993 (Polis and Winemiller, 1996), emphasized dynamic predator-prey models, causes and effects of spatial and temporal variation, life-history strategies, top-down and bottom-up processes, and comparisons of aquatic, terrestrial, and soil webs. Over the last decade, the ecological debate became increasingly dominated by a number of new topics, such as environmental change, spatial ecology, and functional implications of biodiversity. This has changed our view on the entities, scales, and processes that have to be addressed by ecological research, and the food-web approach became recognized as a most powerful tool to approach these issues. This was the point of departure for the third food-web symposium held in November 2003 in Schloss Rauischolzhausen, Germany (de Ruiter et al., 2005).

The volume published by de Ruiter et al. (2005) constitutes the proceedings of the above (2nd) symposium. Much emphasis is laid on the understanding of food-web structure and stability. Some contributions approach food-web structure and dynamics from "outside" environmental variability, in space as well as in time. Other contributions take the opposite approach by looking in depth to the dynamics of populations and biological attributes of individual within populations. Comparison of food-web structures from different habitats, soil, terrestrial, and aquatic, shows regular patterns in the flows with which food is transferred and processed by the trophic groups in the food webs. This approach bridged the gap between looking at descriptive properties of food-web structure, such as species richness and trophic levels and looking at species composition in detail, as it reveals regularities in food web structure that are crucial to food-web stability and functioning and appears less sensitive to the dynamics in species composition in food webs (de Ruiter et al., 2005).

The evolution of realistic food-web structures can be explained on the basis of simple rules regarding population abundance and species occurrence. Life-history-based dynamics within populations may even influence community dynamics in extraordinary and counterintuitive ways in the way that predators promote each other's persistence when they forage on different life stages of their prey, inducing a shift in the size distribution of the prey, leading to more and larger sized individuals and increased population fecundity (de Ruiter et al., 2005).

But also within populations the dynamics in the behavior of individuals, such as prey switching, may affect population dynamics, as dietary shifts inhibit rapid growth by abundant prey and at the same time allow rare prey to rally. If these shifts are fast enough, food-web architecture changes at the same time scale as population dynamics (de Ruiter et al., 2005). This affects food-web structure and stability and may even result in a positive complexity-stability relationship as proposed by Elton some 70 years ago (Elton, 1927; de Ruiter et al., 2005). Preferential feeding by predators may result from prey properties (body size), or from spatial and temporal variability in prey availability. While dietary shifts may be the result of adaptive behavior by the predator, predators may also "induce" defense mechanism in the prey; the dynamics of attack and defense may have strong implications for food-web structure, stability, and functioning. Ratios between predator and prey body sizes generate patterns in the strengths of trophic interactions that enhance food-web stability in a Scottish estuary. This finding confirms the published analysis of the mammal community of the Serengeti, in which predator-prey body size ratios are a primary factor determining predation risks (Sinclair et al., 2003; de Ruiter et al., 2005).

Resource availability and use may govern the structure and functioning of food webs, in turn food-web interactions are the basis of ecosystem processes and govern important pathways in the global cycling of matter, energy, and nutrients. Food-web studies in this way connect the dynamics of populations to the dynamics in ecosystem processes (Section 5 in de Ruiter et al., 2005). The mutual effects between the dynamics of food webs and detritus influences food web structure as well as habitat quality. Variation in the availability of one environmental factor, i.e., nitrogen deposition, affects ecosystem processes like organic matter decomposition nitrogen mineralization, and CO_2 emission through the mediating role of the soil food web (de Ruiter et al., 2005). To fully understand the role of food webs in the energy cycle requires tools to translate resource availability to energy supply necessary for population functioning and persistence. Mechanisms operating within these transitions may vary among resource of the different trophic levels (e.g., primary producers, herbivores, and carnivores). Models that calculate the interplay between processes and food-web structure and functioning have hardly accounted for such dynamics and variations; hence, new ways of modeling these processes are proposed (de Ruiter et al., 2005).

The trophic context of species in food webs may strongly influence the risks of species loss, and the possible consequences of species loss for ecosystem functioning (Section 6 in de Ruiter et al., 2005). A modeling approach shows that in multitrophic level systems, increasing diversity influences plant biomass and productivity in a nonlinear manner. These model results are supported by empirical evidence showing that the consequences of species loss to ecosystem functioning depend on trophic level. And experiments on pond food webs show that the contributions of species to ecosystem processes depend on environmental factors, such as productivity, as well as on trophic position whereby higher trophic levels tend to have the largest effects. These kinds of results indicate that the effects of a particular species loss on ecosystem functioning can be inconsistent across ecosystems. In soil food webs, the role of species in soil processes functional redundancy greater within trophic groups than between trophic groups. Similarly, community invasibility does not entirely depend on factors like resource availability, but also on community structure especially when the "receiving" food web becomes more reticulate. These model and experimental findings ask for new ways to measure functional diversity of species depending on the trophic structure of which they are part of (de Ruiter et al., 2005).

In the field of environmental risk assessment, food webs provide a way to analyze the overall assemblage of direct and indirect effects of environmental stress and disturbance (Section 7 in de Ruiter et al., 2005). Such indirect effects occur through the transfer and magnification of contaminants through food chains causing major effects on species at the end of the food chain, as well as through changes in the dynamics of interacting populations. Sometimes, species extinctions can be seen as the direct result of human activities, but in other cases, extinctions are to be understood from effects of primary extinctions on the structure of the food web, such as the disappearance of some bird species from Barro Colorado Island. Overexploitation by fisheries is one of the most acute environmental problems in freshwater as well as in marine systems. Regarding sustainable fisheries, most ecologists are familiar with the "fishing down food webs" phenomenon (Pauly et al., 1998). The complex nature of effects of human activities on ecosystem properties asks for ways to communicate these effects with resource managers and policymakers.

Visualization in the form of food webs has been shown to be very helpful. In this way, food-web approaches are increasingly providing guidance for the assessment of ecological risks of human activities and for the sustainable management of natural resources and are even beginning to influence policy. The next years of food-web research should see continued theoretical advancement accompanied by rigorous experiments and detailed empirical studies of food-web modules in a variety of ecosystems. Future studies are needed to examine effects of taxonomic, temporal, and spatial scales on dynamic food-web models. For example, adaptive foraging partially determines and stabilizes food-web dynamics, but environmental heterogeneity at appropriate scales also can have a stabilizing effect. A challenge will be to further elaborate the intriguing idea that trophic interactions in food webs drive patterns and dynamics observed at multiple levels of biological organization (Section 1.1 in de Ruiter et al., 2005).

7.2 Food-web science on the path from abstraction to prediction

This section follows the chapter 1.2 of Winemiller and Layman in the book by de Ruiter et al., 2005. The chapter explores some basic issues in food-web research, evaluates major obstacles impeding empirical research, and proposes a research approach aimed at improving predictive models through descriptive and experimental studies of modules within large, complex food webs. Challenges for development of predictive models of dynamics in ecosystems are formidable; nonetheless, much progress has been made during the three decades leading up to this third workshop volume. In many respects, food-web theory has outpaced the empirical research needed to evaluate models. We argue that much greater investment in descriptive and experimental studies and exploration of new approaches are needed to close the gap.

The most fundamental questions in food-web science are "How are food webs structured?" and "How does this structure influence population dynamics and ecosystem processes?" At least four basic models of food web structure can be proposed. One model could be called the "Christmas tree" model, in which production dynamics and ecosystem processes essentially are determined by a relatively small number of structural species. Most of the species' richness in communities pertains to interstitial species that largely depend on the structural species for resources and may be strongly influenced by predation from structural species. Hence, interstitial species are like Christmas ornaments supported by a tree composed of structural species (see Fig. 7.1A, "Christmas tree," of Winemiller and Layman, 2005).

Structural species could include conspicuous species that dominate the biomass of the system, but could also be keystone species that may be uncommon but have disproportionately large effects on the food web and ecosystem (Power et al., 1996; Hurlbert, 1997). In many ecosystems, certain plants and herbivores clearly support most of the consumer biomass, and certain consumers strongly influence biomass and production dynamics at lower levels. This pattern may be more apparent in relatively low-diversity communities, such as shortgrass prairies and kelp forests, in which relatively few species provide most of the production, consume most of the resources, or influence most of the habitat features.

A second alternative is the "onion" model in which core and peripheral species influence each other's dynamics, with core species having a greater influence (i.e., magnitudes of pairwise

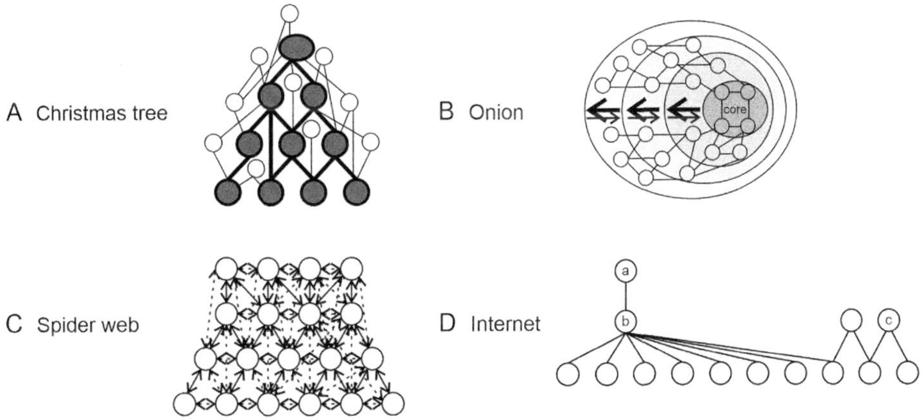

Fig. 7.1
Four basic food-web structures as classified by Winemiller and Layman (2005).

species effects are not reciprocal). The core-peripheral structure is arranged in a nested hierarchy (Fig. 7.1B in Winemiller and Layman, 2005).

This model might pertain to high-diversity ecosystems such as tropical rainforests and coral reefs. Ecological specialization via co-evolution would result in interactions from peripheral species that may have strong effects on a few species, but weak effects on most of the community, and very weak effects on core species. In tropical rainforests, rare epiphytic plants and their co-evolved herbivores, pollinators, and seed dispersers depend upon the core assemblage of tree species, yet the converse is not true. Removal of a given pollinator species would yield a ripple effect within an interactive subset, or module, of the food web, but probably would not significantly affect core species of decomposers, plants, and animals.

A third food-web structure could be called the "spider web" model in which every species affects every other species via the network of direct and indirect pathways (Fig. 7.1C in Winemiller and Layman, 2005).

This concept, in which everything affects everything, is explicit in network analysis (Fath and Patten, 1999), which gives rise to numerous emergent properties of networks (Ulanowicz, 1986). Signal strength, via direct or indirect propagation, may depend on proximity of nodes within the network. Propagation of indirect effects in food webs can yield counterintuitive results from press perturbations. For example, harvesting a competitor of a top predator can result in a decline rather than an increase of that predator (Yodzis, 1996; Wootton, 2001; Relyea and Yurewicz, 2002).

A fourth model of food-web structure could be called the "internet" model. Following this concept, webs are networks having major and minor "hubs" in which their position within the network architecture determines the degree that a species can influence other species in the system (Fig. 7.1D in Winemiller and Layman, 2005).

Jordán and Scheuring (2002) reviewed the applicability of the internet model to food webs and maintained that the density of connections to a node may be a poor indicator of the potential influence on web dynamics. For example, a highly influential species (e.g., top predator) could have only one or a few links connecting it to other species that in turn have numerous connections to other species in the system. Analysis of network features has become a popular pursuit in fields ranging from the social sciences to cell biology, but the relevance of this approach for understanding food-web dynamics is uncertain (Jordán and Scheuring, 2002).

How are food webs structured? The answer will necessarily rely on accumulated evidence from a large body of empirical research. We contend that available evidence is insufficient to state, with a degree of confidence, the general circumstances that yield one or another of these alternative models. Like any scientific endeavor, research on food webs advances on four interacting fronts: description (observation), theory (model formulation), model testing (experimentation), and evaluation. Evaluation invariably leads to theory revision and the loop begins again. After several trips around this loop, a model may begin to successfully predict observations, and we gain confidence for applications to solve practical problems. Important ecological challenges already have been addressed using the food web paradigm, including biocontrol of pests, fisheries management, biodiversity conservation, management of water quality in lakes, and ecotoxicology (Crowder et al., 1996). We believe, however, that the development of food-web theories (models) and their applications is greatly outpacing advances in the descriptive and experimental arenas. Although this state of affairs is not unexpected in an immature scientific discipline, it results in inefficient development of understanding. Why have empirical components lagged behind theoretical developments? We propose that unresolved issues of resolution and scale have hindered empirical research. Resolution of four basic aspects of food webs is required: (1) the food web as an operational unit, (2) components of food webs, (3) the nature of food web links, and (4) drivers of temporal and spatial variation.

7.3 Food webs as units

First, the spatial and temporal boundaries of a community food web are always arbitrary, and it should be emphasized that any food web is a module or subnetwork embedded within a larger system (Cohen, 1978; Moore and Hunt, 1988; Winemiller, 1990; Polis, 1991; Hall and Raffaelli, 1997; Holt, 1977; and others). Food webs are almost always defined according to habitat units nested within, and interacting with larger systems (e.g., biotia living on a single plant, water-filled tree holes, soil, lakes, streams, estuaries, forests, islands). Hence, every empirical food web is a web module. Spatial and taxonomic limits of modules are essentially arbitrary. Thus, it probably makes little sense to speak of large versus small webs, for example. Web modules vary in their degree of correspondence to habitat boundaries.

Although a lake has more discrete physical boundaries than a lowland river with flood pulses and marginal wetlands, numerous links unite lake webs with surrounding terrestrial webs. Thus, broad comparative studies of food-web properties necessarily deal with arbitrary units that may have little or no relationship to each other. To illustrate this point, we examine empirical food webs from three studies, all published in the journal *Nature*, that constructed models to predict statistical features of these webs (Williams and Martinez, 2000; Garlaschelli et al., 2003; Krause et al., 2003). Leaving aside issues related to web links and environmental drivers, let us examine the number of taxa within each habitat. For statistical comparisons, these taxa were subsequently aggregated into “trophospecies” (species that presumably eat all the same resources and also are eaten by all the same consumers). The number of taxa was reported as follows: Skipwith Pond, England (35); Bridge Brook Lake, New York (75); Little Rock Lake, Wisconsin (181); Ythan Estuary, Scotland (92); Chesapeake Bay, United States (33); Coachella Valley, California (30); and Isle of St. Martin, Caribbean (44). Thus, we are led to conclude that Skipwith Pond, a small ephemeral pond in England (Warren, 1989), contains more taxa than

Ythan Estuary, Scotland (92) (Hall and Raffaelli, 1997) and Chesapeake Bay (33) (Baird and Ulanowicz, 1989), one of the world's largest estuaries. These food webs were originally compiled based on different objectives and criteria. The Skipwith Pond food-web reports no primary producer taxa, the Bridge Brook Lake web contains only pelagic taxa, the Chesapeake Bay web is an ecosystem model with a high degree of aggregation, and the Ythan Estuary web includes 27 bird taxa with most other groups highly aggregated. If we examine just the number of reported fish species, Skipwith Pond has none, Ythan Estuary has 17, and Chesapeake Bay is reported to have 12. In reality, Chesapeake Bay has at least 202 fish species (Hildebrand and Schroeder, 1972). These comparative studies analyzed features of Polis's (1991) highly aggregated Coachella Valley web (30 taxa) even though the author clearly cautioned against it and indicated that the web contained, among other taxa, at least 138 vertebrate, 174 vascular plant, and an estimated 2000–3000 insect species. The Isle of St. Martin web was reported to have 44 taxa that include 10 bird and 2 lizard species plus 8 nonvertebrate aggregations (Goldwasser and Roughgarden, 1993). Clearly, these empirical food webs represent an odd collection of woefully incomplete descriptions of community species richness and trophic interactions and are unlikely to provide a basis for robust predictive models.

Discrepancies are due to the fact that these webs were originally compiled based on different objectives and criteria. Objective methods for defining and quantifying nested modules are badly needed. At a minimum, consistent operational definitions for units and standardized methodologies are required to make quantitative comparisons. For example, sink food webs (Cohen, 1978) can be defined based on the network of direct trophic links leading to a predator. Comparisons of different systems could be based on the sink webs associated with predators that are approximate ecological "equivalents." Alternatively, food-web comparisons can be based on the collection of sink webs leading to consumers of a given taxonomic group, such as fishes (Winemiller, 1990). Source webs (tracing the network trophic links derived from a taxon positioned low in the web) provide an operational unit for food web comparisons (e.g., grasses-herbivorous insects-parasitoids) (Martinez et al., 1999), but in most cases, as links radiate upward (to higher trophic positions), they would very rapidly project outward (to adjacent habitats) in a manner that would yield major logistic challenges for empirical study.

7.4 Components of food webs

Our second issue is the units comprising food webs. Entities comprising food webs have been invoked to serve different objectives that are rarely compatible. Consequently, great variation is observed among food-web components, ranging from species life stages to functional groups containing diverse taxa. In most empirical studies, these components have been invoked a posteriori rather than a priori. We must decide a priori whether we wish to examine individuals (what we catch), species populations (what we want to model), "trophospecies" (what we invoke when taxa had been aggregated), functional groups (what we think might be relevant), or trophic levels (what we once thought was relevant). Yodzis and Winemiller (1999) examined multiple criteria and algorithms for aggregating consumers into trophospecies based on detailed abundance and dietary data. Taxa reveal little overlap in resource use and the extent to which predators are shared, and almost no taxa could be grouped according to a strict definition of shared resources and predators.

A similar approach is developed by Luczkovich et al. (2003) in which graph theory and the criterion of structural equivalence are used to estimate degrees of trophic equivalency among taxa. Unlike trophospecies, structurally equivalent taxa do not necessarily feed on any of the same food resources or share even a single predator, but they do play functionally similar roles in the network. Luczkovich et al. (2003) contend that species populations are the only natural food-web components, because populations are evolutionary units with dynamics that are largely independent from those of heterospecific members of a guild or functional group (Ehrlich and Raven, 1969).

7.5 Food-web links

The third issue is how to estimate food-web links. Too often in the past, food-web architecture was treated as binary with links either present or absent (i.e., web topology with no magnitudes or dynamics). Motivated, in part, by the seminal theoretical work of May (1973), empirical studies attempt to determine the nature and magnitude of links (i.e., interaction strength) using field experiments in which one or more species are manipulated (Paine, 1992; Menge, 1995; Wootton, 1997; Raffaelli et al., 2003). Interaction strength determines system dynamics (Paine, 1980) and stability (Yodzis, 1981; Pimm, 1982; McCann et al., 1998), as well as the manner in which we view the basic structure of the food web (de Ruiter et al., 1995; Winemiller, 1990). Weak links are associated with greatest variation in species effects (Berlow, 1999), and food webs seem to be dominated by these weak links.

For example, food webs of tropical aquatic systems are strongly dominated by weak feeding pathways as estimated from volumetric analysis of fish stomach contents (see Figs. 2 and 3 in Winemiller and Layman, 2005). Despite the critical need to understand interaction strength and the manner in which it creates food web structure and drives dynamics, many theoretical and comparative studies that relied on empirical data have not considered species abundances and have portrayed food web links simply as binary. Why has this been the case? First, it is *difficult* to inventory species in natural communities (e.g., Janzen and Hallwachs, 1994). It is *more difficult* to estimate species' relative abundances, even for conspicuous sedentary species like trees (e.g., Hubbell and Foster, 1986; Terborgh et al., 1990). It is *even more difficult* to estimate the presence of feeding relationships (e.g., Thompson and Townsend, 1999). It is yet more difficult to estimate the magnitudes of feeding relationships (Winemiller, 1990; Tavares-Cromar and Williams, 1996). Finally, it is exceedingly difficult to estimate the strength of species interactions (Paine, 1992; Wootton, 1997).

Interaction strength can be inferred indirectly from quantitative dietary analysis, but it is extremely time-consuming and requires a great degree of taxonomic and modeling expertise. The method is not viable for many consumer taxa, because most food items contained in the gut are degraded. Moreover, large samples are needed to estimate diet breadth (i.e., links) accurately and precisely and to reveal important spatial and temporal variation in feeding relationships (Winemiller, 1990). As sample size is increased from 1 to 20 individuals, the mean diet breadth of an omnivorous characid fish from Caño Maraca increases from 2.8 to 3.9, and the average number of feeding links increases from 3.7 to 22 (Fig. 3 in Winemiller and Layman, 2005). Similarly, sampling effort has been shown to affect food-web properties associated with the number of nodes (Bersier et al., 1999). Quantitative estimates of diet composition must be converted to consumption rates for use in dynamic food web models (see Chapter 7.3 by Koen-Alonso and Yodzis, 2005 in the book by de Ruiter et al., 2005).

Interaction strength can be directly estimated via field experiments, but this method is beset with its own set of challenges (Berlow et al., 2004). A major problem is the quantitative measure used to quantify interaction strength. Several indices have been employed (reviewed by Berlow et al., 1999), including a raw difference measure (N-D)/Y; Paine's index (N-D)/DY; community importance (N-D)/Npy; and a dynamic index (ln(N/D))/Yt, in which N=prey abundance with predator present, D=prey abundance with predator absent, Y=predator abundance, p=predator proportional abundance, and t=time. Different indices computed from the same set of experimental data yield very different conclusions (Berlow et al., 1999).

Even if we could agree on a single empirical measure of interaction strength, we would still face serious challenges in estimating community dynamics with this information (Berlow et al., 2004). This is because species interactions typically are nonlinear, which implies that single estimates of interaction strength will be unlikely to assist in building dynamic community models (Abrams, 2001). According to Abrams, "Measuring interactions should mean determining the functional form of per capita growth rate functions, not trying to encapsulate those complicated relationships by a single number." Application of simple models to predict features and dynamics of complex systems would be justified if these models could, a priori, yield successful predictions. Clearly, considerable theoretical and empirical research remains to be done on the crucial issue of interaction strength.

An additional consideration is that food-web links are usually assumed to be consumer resource; however, other kinds of species interactions, such as mutualism and other forms of facilitation, can be critical (Bruno et al., 2003; Berlow et al., 2004). Describing the functional forms of these relationships could be even more challenging. Some of the most important community interactions are not determined by resource consumption. Gilbert (1980) described ecological relationships in a food-web module within a Costa Rican rainforest. This module is delimited by 36 plant species in 6 higher taxa inhabiting 3 habitat types. Each plant species has a set of generalist and specialist herbivores, pollinators, and seed dispersers, some of which are shared with other plants within the module and, in some cases, plants outside the module. In this food web, some of the most critical interactions determining species' abundances and distributions are mutualisms.

7.6 Drivers of temporal and spatial variation

The fourth critical issue is the influence of environmental and life-history variation on food-web structure, species interactions, and population dynamics. Do food-web dynamics drive species abundance patterns, or do species abundance patterns drive food-web dynamics? Species' relative abundances determine functional responses, adaptive foraging, predator switching, and their effects on numerical responses.

Does food-web structure determine relative abundance patterns, or are other factors equally or more important? Interaction strength varies in space and time, sometimes as a function of behavior, but sometimes as a function of environmental variation and species life histories that affect abundance patterns (Polis and Winemiller, 1996). Species with different life histories and ecophysiological adaptations respond differently to environmental variation (Winemiller, 1989a). Species with short generation times and rapid life cycles respond faster to environmental variation (including resource availability) than species with slower life cycles that often reveal large variation in recruitment dynamics and demographic storage effects (Polis and Winemiller, 1996; Scharler et al., 2005 Chapter 8.3 in the book by de Ruiter et al., 2005). Empirical studies have demonstrated how species' abundances and web links change in response to environmental drivers. Rainfall and leaf litter deposition determine food-web patterns in tree holes in tropical Australia (Kitching, 1987). Temporal dynamics in rocky intertidal webs are influenced by local disturbances (Menge and Sutherland, 1987) and coastal currents (Menge et al., 2003). Food webs of streams and rivers vary in relation to seasonal changes in photoperiod and temperature (Thompson and Townsend, 1999) and hydrology (Winemiller, 1990; Marks et al., 2000).

7.7 Theories, tests, and applications

So where are we now? Theory and attempts at application of theory seem to have outpaced observation and model testing. There is little agreement and consistency regarding the use of operational units, methods for quantifying links, indices of interaction strength, etc. The use of confidence intervals is virtually non-existent in empirical food-web research. This state of affairs is perhaps a natural consequence of an "immature" scientific discipline (i.e., abstract concepts, lack of consensus, and empirical rigor). Nonetheless, society demands that ecological science address current problems. Currently, food-web models have low predictive power and certainly lack the precision and accuracy of physical models that allow engineers to put a spaceship on the moon or build a sturdy suspension bridge. Food-web models currently used for natural resource management are highly aggregated and employ crude quantitative estimates of production dynamics and species interactions. Output from these models can be considered educated guesses, yet, currently, we have no other options. It is unreasonable to expect individual investigators or labs to achieve predictive food-web models, yet few are lobbying for empirical food-web research on a grand scale. This state of affairs may be an unfortunate legacy of the IBP (International Biological Program, supported in the 1960–1970s by large sums of national and international science funding aimed at understanding major ecosystems of the planet).

Were past efforts to describe large food webs misguided? Nearly 20 years ago, Winemiller, attempted to describe food webs of tropical streams in a standardized manner based on intensive sampling (Winemiller, 1989b, 1990, 1996). Two continuous years of field research yielded over 60,000 fish specimens and countless invertebrates. Two additional years of laboratory research (19,290 stomachs analyzed) produced data that supported analyses that have been ongoing for 17 years. These quantitative food webs have provided insights into how environmental variation driven by seasonal hydrology affects population dynamics and interactions. Yet, as descriptions of community food webs, these webs suffer from the same limitations that plague other webs. The many issues, both conceptual and methodological, requiring resolution in order to make meaningful comparisons of web patterns ended up being a major discussion topic (Winemiller, 1989b, 1990).

Is there a better way? Winemiller and Layman (2005) advocate a multi-faceted empirical approach for field studies as a means to advance understanding of food webs. Researchers investigating large, complex systems would be better served to investigate food-web modules in a hierarchical fashion. Long-term research mindful of environmental drivers is extremely valuable in this context. Research that blends together description and experimentation will yield models that can then be tested within relevant domains. This approach obviously will require research teams with specialists that collectively provide a range of methodological and taxonomic expertise. Several groups around the world have already adopted this long-term, team research approach to investigate food webs of ecosystems ranging from estuaries (Raffaelli and Hall, 1992) to rainforests (Reagan and Waide, 1996). Winemiller and Layman have attempted this hierarchical modular approach in our research on the Cinaruco River, a floodplain river in the Llanos region of Venezuela. This group is describing nutrient dynamics, primary production, community structure, habitat associations, and feeding interactions in channel and aquatic floodplain habitats during various phases of the annual hydrological cycle in this diverse food web (see Layman et al., Chapter 7.4 in the book by de Ruiter et al. (2005)).

Population abundance and distribution patterns are assessed from field surveys (Jepsen et al., 1997; Arrington and Winemiller, 2003; Hoeinghaus et al., 2003a; Layman and Winemiller, 2004), and feeding links are investigated using dietary and stable isotope analyses (Jepsen et al., 1997; Jepsen and Winemiller, 2002; Winemiller and Jepsen, 2004; see Layman et al.

(2005) Chapter 7.4 in the book by de Ruiter et al., 2005). Layman et al. (2005) also are investigating three food-web modules: (1) benthivorous fishes, benthic biota, detritus, and nutrients; (2) herbivorous fishes interacting with terrestrial and aquatic vegetation; and (3) piscivores and their diverse prey (see again Layman et al. in Chapter 7.4 in the book by de Ruiter et al. (2005)). Field experiments (enclosures, exclosures, and artificial habitats) have been conducted over variable spatial scales in different seasons and habitats to examine species effects on prey assemblages (Layman and Winemiller, 2004) and benthic primary production and particulate organic matter (Winemiller et al., 2006). In virtually all experiments designed to test for top-down effects, one or a small number of fish species (including large detritivores and piscivores) reveal strong and disproportionate effects in this species-rich food web (more than 260 fish species documented). The descriptive research elements have led to creation of models that predict effects of abiotic ecosystem drivers (the most fundamental being seasonal hydrology) and aspects of species life histories (e.g., seasonal migration by a dominant benthivorous species) on food-web dynamics and ecosystem processes. For example, the relative influence of top-down and bottom-up processes on benthic primary production, benthic particulate organic matter, and meiofauna diversity is a function of the seasonal cycle of hydrology, habitat volume, allochthonous nutrient inputs, migration by the dominant benthivorous fish, and changing densities of resident benthivorous fishes as a function of habitat volume. Experiments have been conducted to estimate the magnitude of treatment effects that reveal the relative influence of bottom-up (nutrient limitation and sedimentation) and top-down (grazing) effects on standing stocks of algae and fine particulate organic matter (Winemiller et al., 2006). A separate series of experiments examined effects of predators on prey fish densities and habitat use (Layman and Winemiller, 2004). The relative influence of dominant piscivores on littoral zone fish assemblages is strongly dependent on body size relationships (see Layman et al., Chapter 7.4 in the book by de Ruiter et al. (2005).) and habitat features which in turn are influenced by seasonal hydrology. In short, almost no aspect of this river food web could be understood without examining the direct and indirect effects of the annual hydrological cycle.

Guided by the descriptions of the overall food web, the predictive models developed for modules are being joined together based on elements of overlap. The degree to which predictions of module dynamics will agree with predictions from a model that incorporates all elements remains to be investigated. Nonetheless, it seems more rational to begin at smaller scales and work incrementally toward a model of the larger system, rather than the reverse approach. A few contributions cited in this book describe similar small-to-large approaches employing multiple lines of empirical evidence to test model predictions.

Several research groups have reported results from long-term research that blends description, experimentation, and modeling—for example, temperate lakes (Carpenter and Kitchell, 1993; Tittel et al., 2003), soils (de Ruiter et al., 1995; Moore et al., 2003), coastal systems (Menge et al., 2003; Raffaelli et al., 2003), rivers and streams (Marks et al., 2000; Nakano and

Murakami, 2001; Flecker et al., 2002), ponds (Downing and Leibold, 2002), and fields (Schmitz, 2003). It is still too early to generalize about food-web structure, and perhaps some systems conform to the onion model, whereas others function according to the internet model, and so on. Given the disproportionate effects of a few dominant species demonstrated by field experiments in the Cinaruco River, the "Christmas tree" and "internet" models seem to be candidates for that species-rich system.

7.8 Discussion and conclusions

Empirical food-web research lags behind theoretical research. Winemiller and Layman et al. (2005) agree with Englund and Moen's (2003) assertion "that it is vital for progress in ecology that more models are experimentally tested, and the main question is how to promote and speed up the process." They continue: "By testing a model, we mean the act of comparing model predictions with relevant empirical data." Another basic challenge identified by these authors is the critical need to determine whether or not an experimental system lies within the theoretical domain of the model being tested. In too many cases, models and tests were mismatched from the start (e.g., invalid assumptions of linear or equilibrium dynamics or inappropriate spatial scales). Empirical food-web studies must carefully consider the dynamical consequences of definitions for operational units and scale, resolution, and sample variability. Obviously, it is impossible to quantify every species and interaction in even the smallest food-web modules. Even if this were possible, it is unlikely that most trophic interactions have a strong effect on system properties such as nutrient cycling and production of dominant biomass elements (e.g., the "Christmas tree" and "internet" models). Thus, it is crucial that we determine, to the possible extent, the degree of resolution needed to make successful predictions, and then, for the sake of efficiency, not seek to achieve high levels of detail for their own sake. We advocate a focus on a hierarchy of nested food-web modules and measures of interaction strength that hold potential to yield successful predictions of population dynamics and other ecosystem features. Descriptive and experimental research should be combined in long-term studies of field sites (see also Schmitz, 2001). Such efforts require consistent funding and collaborations among scientists with different expertise. In many countries, these sorts of projects are difficult to fund and provide fewer individual rewards than short-term projects addressing specific mechanisms in small-scale ecological systems. Yet, many of our most vexing ecological problems require a large-scale food-web perspective. Despite the fact that a deficient empirical knowledge base is the main hurdle to scientific advancement, pressing natural resource problems require application of existing models. In many respects, food-web research is basic yet complicated—esoteric yet essential for natural resource management. The urgent need for application of the food-web paradigm for solving natural resource problems motivates us to walk faster down the path from abstraction to prediction (Winemiller and Layman, 2005).

References

Abrams, P.A., 2001. Describing and quantifying interspecific interactions: a commentary on recent approaches. Oikos 94, 209–218.

Arrington, D.A., Winemiller, K.O., 2003. Diel changeover in sandbank fish assemblages in a Neotropical floodplain river. J. Fish Biol. 63, 442–459.

Baird, D., Ulanowicz, R.E., 1989. The seasonal dynamics of the Chesapeake Bay ecosystem. Ecol. Monogr. 59, 329–364.

Berlow, E.L., 1999. Strong effects of weak interactions in ecological communities. Nature 398, 330–334.

Berlow, E.L., Navarrete, A.A., Briggs, C.J., Power, M.E., Menge, B.A., 1999. Quantifying variation in the strengths of species interactions. Ecology 80, 2206–2224.

Berlow, E.L., Neutel, A.M., Cohen, J.E., De Ruiter, P.C., Ebenman, B., Emmerson, M., Fox, J.W., Jansen, V.A.A., Jones, J.I., Kokkoris, G.D., Logofet, D.O., McKane, A.J., Montoya, J.M., Petchey, O., 2004. Interaction strengths in food webs: issues and opportunities. J. Anim. Ecol. 73 (3), 585–598.

Bersier, L.-F., Dixon, P., Sugihara, G., 1999. Scale-invariant or scale-dependent behavior of link density property in food webs: a matter of sampling effort? Am. Nat. 153 (6), 676–682.

Bruno, J.F., Stachowicz, J.J., Bertness, M.D., 2003. Inclusion of facilitation into ecological theory. Trends Ecol. Evol. 18, 119–125.

Carpenter, S.J., Kitchell, J.F., 1993. The Trophic Cascade in Lakes. Cambridge University Press, Cambridge, p. 385.

Cohen, J.E., 1978. Food Webs in Niche Space. Princeton University Press, Princeton, NJ.

Crowder, L.B., Reagan, D.P., Freckman, D.W., 1996. Food web dynamics and applied problems. In: Polis, G.A., Winemiller, K.O. (Eds.), Food Webs: Integration of Patterns and Dynamics. Chapman and Hall, New York, pp. 327–336.

de Ruiter, P.C., Moore, J.C., 2004. Trophic interactions and the dynamics and stability of soil populations and food webs. In: Hillel, D. (Ed.), Encyclopedia of Soils in the Environment. Academic Press, ISBN: 9780123485304.

de Ruiter, P.C., Neutel, A.M., Moore, J.C., 1995. Energetics patterns of interaction strengths and stability in real ecosystems. Science 269, 1257–1260.

de Ruiter, P.C., Wolters, V., Moore, J.C., 2005. Dynamic Food Webs, sec. 1.1: Multispecies, Assemblages, Ecosystem Development and Environmental Change., ISBN: 978-0-12-088458-2, pp. 3–9 (Cited occasionally as de Ruiter et al. (2005)).

DeAngelis, D.L., 1992. Dynamics in Food Webs and Nutrient Cycling. Chapman & Hall, London, UK.

DeAngelis, D.L., Post, W., Sugihara, G., 1982. Current Trends in Food Web Theory. 5983. Oak Ridge National Laboratory, Oak Ridge, TN.

Downing, A.L., Leibold, M.A., 2002. Ecosystem consequences of species richness and composition in pond food webs. Nature 416, 837–841.

Ehrlich, P.R., Raven, P.H., 1969. Differentiation of populations. Science 165, 1228–1232.

Elton, C.S., 1927. Animal Ecology. Sidgwick and Jackson, London.

Englund, G., Moen, J., 2003. Testing models of trophic dynamics: the problem of translating from model to nature. Aust. J. Ecol. 28, 61–69.

Fath, B.D., Patten, B.C., 1999. Review of the foundations of network environ analysis. Ecosystems 2, 167–179.

Flecker, A.S., Taylor, B.W., Berhardt, E.S., Hood, J.M., Cornwell, W.K., Cassatt, S.R., Vanni, M.J., Altman, N.S., 2002. Interactions between herbivorous fishes and limiting nutrients in a tropical stream ecosystem. Ecology 83, 1831–1844.

Garlaschelli, D., Caldarelli, G., Pietronero, L., 2003. Universal scaling relations in food webs. Nature 423, 165–168.

Gilbert, L.E., 1980. Food web organization and the conservation of neotropical diversity. In: Soulé, M.E., Wilcox, B.A. (Eds.), Conservation Biology: An Evolutionary Perspective. Sinauer, Sunderland, Massachusetts, pp. 11–33.

Goldwasser, L., Roughgarden, J., 1993. Construction and analysis of a large Caribbean food web. Ecology 74, 1216–1233.

Hall, S.J., Raffaelli, D., 1997. Food web patterns: what do we really know? In: Gange, A., Brown, A.C. (Eds.), Multitrophic Interactions in Terrestrial Systems. Blackwell University Press.

Hildebrand, S.F., Schroeder, W.C., 1972. Fishes of Chesapeake Bay. T.F.H, Neptune, New Jersey, p. 388.

Hoeinghaus, D.J., Layman, C.A., Arrington, D.A., Winemiller, K.O., 2003a. Spatiotemporal variation in fish assemblage structure in tropical floodplain creeks. Environ. Biol. Fish 67, 379–387.

Holt, R.D., 1977. Predation, apparent competition, and the structure of prey communities. Theor. Popul. Biol. 12, 197–229.

Hubbell, S.P., Foster, R.B., 1986. Commonness and rarity in a neotropical forest: implications for tropical tree conservation. In: Soulé, M. (Ed.), Conservation Biology: Science of Scarcity and Diversity. Sinauer Associates, Sunderland, Massachusetts, pp. 205–231.

Hurlbert, S.H., 1997. Functional importance vs. keystoneness: reformulating some questions in theoretical biocenology. Aust. J. Ecol. 22, 369–382.

Janzen, D.H., Hallwachs, W., 1994. All Taxa Biodiversity Inventory (ATBI) of Terrestrial Systems. A Generic Protocol for Preparing Wildland Biodiversity for Non-Damaging Use. Report of a NSF Workshop, Philadelphia, Pennsylvania.

Jepsen, D.B., Winemiller, K.O., 2002. Structure of tropical river food webs revealed by stable isotope ratios. Oikos 96, 46–55.

Jepsen, D.B., Winemiller, K.O., Taphorn, D.C., 1997. Temporal patterns of resource partitioning among Cichla species in a Venezuelan black-water river. J. Fish Biol. 51, 1085–1108.

Jordán, F., Scheuring, I., 2002. Searching for keystones in ecological networks. Oikos 99, 607–612.

Kitching, R.L., 1987. Spatial and temporal variation in food webs in water-filled treeholes. Oikos 48, 280–288.

Koen-Alonso, M., Yodzis, P., 2005. Dealing with model uncertainty in trophodynamic models: a patagonian example, Sec. 7.3. In: de Ruiter, P.C., Wolters, V., Moore, J.C. (Eds.), Dynamic Food Webs. Elsevier–Academic Press, pp. 381–394.

Krause, A.E., Frank, K.A., Mason, D.M., Ulanowicz, R.E., Taylor, W.W., 2003. Compartments revealed in food-web structure. Nature 426, 282–285.

Layman, C.A., Winemiller, K.O., 2004. Size-based response of multiple species to piscivore exclusion in a Neotropical river. Ecology 85, 1311–1320.

Layman, C.A., Winemiller, K.O., Arrington, D.A., 2005. Patterns of habitat segregation among large fishes in Venezuelan floodplain river. Neo. Ichthyol. 3, 111–117 (Chapter 7.4 in the book by de Ruiter et al. (2005)).

Luczkovich, J.J., Borgatti, S.P., Johnson, J.C., Everett, M.G., 2003. Defining and measuring trophic role similarity in food webs using regular equivalence. J. Theor. Biol. 220, 303–321.

Marks, J.C., Power, M.E., Parker, M.S., 2000. Flood disturbance, algal productivity, and interannual variation in food chain length. Oikos 90, 20–27.

Martinez, N.D., Hawkins, B.A., Dawah, H.A., Faiferek, B.P., 1999. Effect of sampling effort on characterization of food-web structure. Ecology 80 (3), 1044–1055.

May, R.M., 1972. Will a large complex system be stable? Nature 238, 413–414.

May, R.M., 1973. Stability and Complexity in Model Ecosystems, second ed. Princeton University Press, Princeton, USA.

McCann, K., Hastings, A., Huxel, G.R., 1998. Weak trophic interactions and the balance of nature. Nature 395, 794–798.

Menge, B.A., 1995. Indirect effects in rocky intertidal interaction webs: patterns and importance. Ecol. Monogr. 65, 21–74.

Menge, B.A., Lubchenco, J., Bracken, M.E.S., Chan, F., Foley, M.M., Freidenburg, T.L., Gaines, S.D., Hudson, G., Krenz, C., Menge, D.N.L., Russell, R., Webster, M.S., 2003. Coastal oceanography sets the pace of rocky intertidal community dynamics. Proc. Natl. Acad. Sci. U. S. A. 100, 12229–12234.

Menge, B.A., Sutherland, J.P., 1987. Community regulation: variation in disturbance, competition, and predation in relation to environmental stress and recruitment. Am. Nat. 130, 730–757.

Moore, J.C., Hunt, H.W., 1988. Resource compartmentation and the stability of real ecosystems. Nature 333, 261–263.

Moore, J.C., McCann, K., Setälä, H., de Ruiter, P., 2003. Top-down is bottom-up: does predation in the rhizosphere regulate aboveground dynamics? Ecology 84, 846–857.
Nakano, S., Murakami, M., 2001. Reciprocal subsidies: dynamic interdependencebetween terrestrial and aquatic food webs. Proc. Natl. Acad. Sci. U. S. A. 98, 166–170.
Odum, E.P., 1963. Ecology. Holt, Rinehard, and Winston, New York.
Paine, R.T., 1980. Food webs: linkage, introduction, strength and community infrastructure. J. Anim. Ecol. 4913, 666–685.
Paine, R.T., 1992. Food-web analysis through field measurements of per capita interaction strength. Nature 355, 73–75.
Pauly, D., Christensen, V., Dalsgaard, J., Froese, R., Torres, F., 1998. Fishing down marine food webs. Science 279, 860–863.
Pimm, S.L., 1982. Food Webs. Chapman and Hall, London.
Polis, G.A., 1991. Complex trophic interactions in deserts: an empirical critique of food web theory. Am. Nat. 138, 123–155.
Polis, G.A., Winemiller, K.O. (Eds.), 1996. Food Webs: Integration of Patterns and Dynamics. Chapman & Hall, New York.
Power, M.E., Tilman, D., Estes, J.A., Menge, B.A., Bond, W.J., Mills, L.S., Daily, G., Castilla, J.C., Lubchenco, J., Paine, R.T., 1996. Challenges in the quest for keystones. Bioscience 46, 609–620.
Raffaelli, D., Emmerson, M., Solan, M., Biles, C., Paterson, D., 2003. Biodiversity and ecosystem processes in shallow coastal waters: an experimental approach. J. Sea Res. 49, 133–141.
Raffaelli, D., Hall, S.J., 1992. Compartmentation and predation in an estuarine food web. J. Anim. Ecol. 61, 551–560.
Reagan, D.P., Waide, R.B. (Eds.), 1996. The Food Web of a Tropical Rain Forest. The University of Chicago Press, Chicago.
Relyea, R.A., Yurewicz, K.L., 2002. Predicting community outcomes from pairwise interactions: integrating density- and trait-mediated effects. Oecologia 131, 569–579.
Scharler, U.M., Hulot, F.D., Baird, D.J., Cross, W.F., Culp, J.M., Layman, C.A., Raffaelli, D., Vos, M., Winemiller, K.O., 2005. Central Issues for Aquatic Food Webs: From Chemical Cues to Whole System Responses., p. 451 (Chapter 8.3 in the book by de Ruiter et al., 2005).
Schmitz, O.J., 2001. From interesting detail to dynamic relevance: toward more effective use of empirical insights in theory construction. Oikos 94, 39–50.
Schmitz, O.J., 2003. Top predator control of plant biodiversity and productivity in an old field ecosystem. Ecol. Lett. 6, 156–163.
Sinclair, A.R.E., Mduma, S., Brashares, J.S., 2003. Patterns of predation in a diverse predator-prey system. Nature 425, 288–290.
Tavares-Cromar, A.F., Williams, D.D., 1996. The importance of temporal resolution in food web analysis: evidence from a detritus-based stream. Ecol. Monogr. 66, 91–113.
Terborgh, J., Robinson, S.K., Parker III, T.A., Munn, C.A., Pieront, N., 1990. Structure and organization of an Amazonian forest bird community. Ecol. Monogr. 60, 213–238.
Thompson, R.M., Townsend, C.R., 1999. Is resolution the solution?: the effect of taxonomic resolution on the calculated properties of three stream food webs. Freshw. Biol. 44, 413–422.
Tittel, J., Bissinger, V., Zippel, B., Gaedke, U., Bell, E., Lorke, A., Kamjunke, N., 2003. Mixotrophs combine resource use to outcompete specialists: implications for aquatic food webs. Proc. Natl. Acad. Sci. U. S.A. 100, 12776–12781.
Ulanowicz, R.E., 1986. Growth and Development: Ecosystems Phenomenolgy. Springer-Verlag, New York, p. 232.
Warren, P.H., 1989. Spatial and temporal variation in the structure of a freshwater food web. Oikos 55, 299–311.
Williams, R.J., Martinez, N.D., 2000. Simple rules yield complex food webs. Nature 404, 180–183.
Winemiller, K.O., 1989a. Patterns of variation in life history among South American fishes in seasonal environments. Oecologia 81, 225–241.
Winemiller, K.O., 1989b. Must connectance decrease with species richness? Am. Nat. 34, 960–968.

Winemiller, K.O., 1990. Spatial and temporal variation in tropical fish trophic networks. Ecol. Monogr. 60, 331–367.
Winemiller, K.O., 1996. Factors driving spatial and temporal variation in aquatic floodplain food webs. In: Polis, G.A., Winemiller, K.O. (Eds.), Food Webs: Integration of Patterns and Dynamics. Chapman and Hall, New York, pp. 298–312.
Winemiller, K.O., Jepsen, D.B., 2004. Migratory neotropical fish subsidize food webs of oligotrophic Blackwater rivers. In: Polis, G.A., Power, M.E., Huxel, G. (Eds.), Food Webs at the Landscape Level. University of Chicago Press, Chicago, pp. 115–132.
Winemiller, K.O., Layman, C.A., 2005. In: de Ruiter, P.C., Wolters, V., Moore, J.C. (Eds.), Dynamic Food Webs, Sec. 1.2: Food Web Science: Moving on the Path from Abstraction to Prediction, pp. 10–23 (last page of Sec. 1.2).
Winemiller, K.O., Montoya, J.V., Layman, C.A., Roelke, D.L., Cotner, J.B., 2006. Experimental demonstrations of seasonal fish effects on benthic ecology of a neotropical floodplain river. J. N. Am. Benthol. Soc. 25 (1), 250–262.
Wootton, J.T., 1997. Estimates and tests of per capita interaction strength: diet, abundance, and impact of intertidally foraging birds. Ecol. Monogr. 67, 45–64.
Wootton, J.T., 2001. Prediction in complex communities: analysis of empirically derived Markov models. Ecology 82, 580–598.
Yodzis, P., 1981. The stability of real ecosystems. Nature 289, 674–676.
Yodzis, P., 1996. Food webs and perturbation experiments: theory and practice. In: Polis, G.A., Winemiller, K.O. (Eds.), Food Webs: Integration of Pattern and Dynamics. Chapman and Hall, New York, pp. 192–200.
Yodzis, P., Winemiller, K.O., 1999. In search of operational trophospecies in a tropical aquatic food web. Oikos 87, 327–340.

Further reading

Hoeinghaus, D.J., Layman, C.A., Arrington, D.A., Winemiller, K.O., 2003b. Movement of Cichla spp. (Cichlidae) in a Venezuelan floodplain river. Neotrop. Ichthyol. 1, 121–126.
Werner, E.E., 1998. Ecological experiments and a research program in community ecology. In: Restarits, J.W.J., Bernardo, J. (Eds.), Experimental Ecology: Issues and Perspectives. Oxford University Press, Oxford, pp. 3–26.

CHAPTER 8

Outline of mathematical ecology by E.C. Pielou

8.1 Pielou's preface and introduction

The title of 1969 Wiley's book authored by E.C. Pielou is: An Introduction to Mathematical Ecology. The present author (S.S.) takes the liberty of citing from the book's several representative sentences which are quoted in the text below. As stressed by E.C.P in her Preface, "The fact that ecology is essentially a mathematical subject is becoming ever more widely accepted. Ecologists everywhere are attempting to formulate and solve their problems by mathematical reasoning, using whatever mathematical knowledge they have acquired, usually in undergraduate courses or private study. The purpose of the book is to serve as a text for these students and to demonstrate the wide array of ecological problems that invite continued investigation." Below are samples of other citations from the same Preface:

> In writing a book of this length (or, indeed, of any length), the author is always faced with the problem whether to cover a great many topics superficially or to delve deeply into a few. The compromise I have attempted has been to pick selected topics over the whole range of mathematical ecology and then to deal in detail with those aspects that seen likely to furnish good starting points for further research. The list of chapter headings shows the subjects that were chosen. Their choice and the aspects to be preserved, has, of course, been subjective and a matter of my own judgment. It is unlikely that any other ecologist with the same objective would have selected exactly the same topics.
>
> ***(E.C.P stressing)***

Next Pielou states:

> The book is in no sense a review. An enormous account of interesting and valuable work has been ignored and any mathematical ecologist glancing at the bibliography is bound to wonder at what may see to him inexplicable omissions. The decision to exclude an account of an interesting piece of work was always difficult, but the exclusions were necessary to make room for a fairly full exposition of the matters dealt with. My object has been to provide a sufficiently detailed development of the topics examined for the reader to be able to consult the current literature with an understanding of what has lead up to it; as far as possible nothing has been asserted without proof. For each topic I have attempted to give a connected account of the underlying theory. So that the reader will not lose the thread of an argument,

Complexity and Complex Ecological Systems. https://doi.org/10.1016/B978-0-443-19237-1.00005-3

> methods of estimating parameters and similar practical details have not been given. Information on these matters, and numerical examples will be found in the literature cited. To make the text available for private study, many mathematical derivations have been written out *in extenso*. I have striven to avoid the mathematician's daunting phrases [...] which are so often substituted for long chains of reasoning that, for the mathematically inexperienced, are neither obvious, nor clear. When these phrases are used here, a line or two at most of algebraic manipulations is all that the reader needs to interpolate.

Now the time comes for the reader to get acquainted with the contents of Pielou's book. Then, the reader moves on to her Introduction on page 1, the content of which is briefly summarized below. The Introduction ensures the reader that the reading will be a difficult matter, because "in an acre of forest, for instance, an enormous number of species is present, from trees to soil microorganisms. Not only does each species differ from every other, but also all the individuals within any one species are unique. Each individual is a complex organism that changes continuously; at any moment its behavior depends on its genetic constitution that prevails in its locality at that moment[...]. In addition to all this, a community as a whole is at all times being depleted by deaths and replenished by births; often immigration and emigration are further causes of a continual turnover of individuals. Thus the components of a community are never the same on two successive occasions [...]." After the specification of many other difficulties, E.C.P. terminates her Introduction by the following sentences: "An introduction necessarily deals with broad generalities. One should not, however, draw the conclusion that mathematical ecology, as a subject, is a unified whole; far from it. Anyone who is dismayed by the seemingly fragmentary nature of the work described in this book must console himself with the thought that the subject is still in its early stages and that the welding together of its disconnected parts is a challenging job yet to be done [ ]." The sentence terminated by the phrase about "a challenging job yet to be done [ ]" ends the Introduction and facilitates the passage to the first chapter titled *Population Dynamics*; see the present author's brief comments below.

8.2 Population dynamics

Chapter 1 on population dynamics begins considerations of birth and death processes. The book's author, E.C. Pielou, uses exact terminology and applies a rigorous mathematical formalism. The considered problems are briefly characterized in Sections 8.2.1–8.5.5 of the book. Mainly, simple verbal characteristics are given below. Some fresh ideas in ecology as, e.g., ascendant perspective of Ulanowicz (1997) are too new to appear in this 1969 book.

8.2.1 Birth and death processes

Since all populations of organisms fluctuate in size, the only assertion is that the size will not remain constant. The investigation of the growth and deadline of populations leads to a simple model to account for population change and then to examining the consequences of the accepted assumptions by mathematical argument. The author therefore begins by considering the simplest, the pure birth process. Then, she deals with the simple birth and death process. The associated mathematics is developed (Pielou, 1969) which terminates with determining the chance of the population extinction.

In the pure birth process, the assumptions are as follows: The organisms are assumed to be immortal and to reproduce at the rate which is the same for every individual and does not change in time. It is also assumed that individuals have no effect on one another. Since no deaths occur, a population growing in this manner can only increase or remain constant; it cannot decrease. In spite of the extreme simplicity of these assumptions, they may apply, approximately at least and over a short time interval, to the growth of a population of a single- celled organisms that reproduce by dividing. Possibly, algal blooms in eutrophic lakes increase in this manner in spring.

If we write N_i for the size of the population at time t and λ for the rate of increase in each individual, it follows that

$$dN_i/dt = \lambda N_i \tag{8.1}$$

hence, ln $N_i = \lambda t + C$, where C is an integration constant. Assume that the initial size of the population at time t was i. Then, C may be evaluated, since at $t=0$ and ln $i = C$. Therefore,

$$\ln N_i/i = \lambda t \tag{8.2}$$

or

$$N_i = i \exp(\lambda t) \tag{8.3}$$

This is the Malthusian equation for the population growth. It shows that the growth is exponential in the simple circumstances assumed. The deterministic process assumes not that an organism may reproduce but that in fact it does reproduce with absolute certainty. Clearly, however, population growth is a stochastic process. Given a population of yeast cells growing by fission, for example, one can say only that there is a certain probability that a particular cell will divide in a given time interval. We must therefore investigate the stochastic form of the pure birth process. The reader is referred to Pielou (1969) book for this approach, more details, and the related stochastic generalization of the deterministic conclusion. This stochastic generalization is, in fact, a complex differential-difference equation for the probability of a birth in the population of size N. Starting from this equation, we can find the probability that a cell will divide and the probability of a birth in the population in terms of the variables λ, N, t, and the initial size of the population at time $t=0$. The form or the general solution for the probability of a pure birth is:

$$p_N(t) = \binom{N-1}{i-1} e^{-\lambda i t}\left(1 - e^{-\lambda t}\right)^{N-i} \tag{8.4}$$

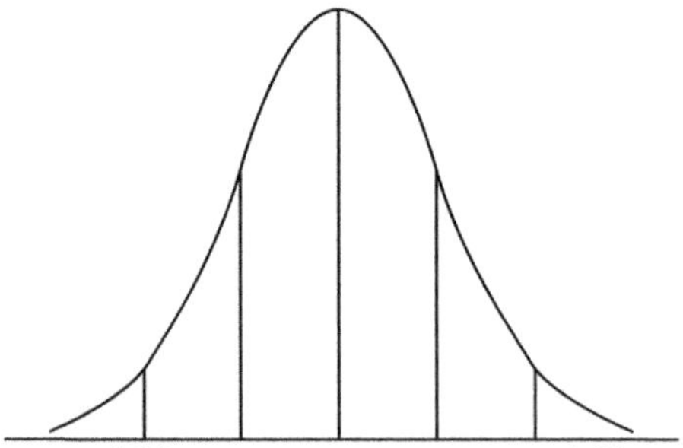

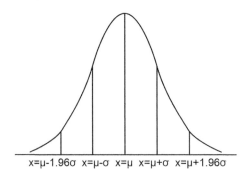

Fig. 8.1
The probability distribution at time t of a population undergoing a pure birth process.

Fig. 1 on p. 11 in Pielou (1969) book shows the probability distribution of the size N at time t of a population undergoing a pure birth process. For the reader's convenience, the draft of this figure is repeated here as our Fig. 8.1.

$$p_N(t) = \binom{N-1}{i-1} e^{-\lambda it} \left(1 - e^{-\lambda t}\right)^{N-i}$$

with $\lambda t = 0.5$ and $i = 5$. The mean is $M(N \mid t) = ie^{\lambda t} = 8.24$. The variance is var.$(N \mid t) = ie^{\lambda t}$ $(e^{\lambda t} - 1) = 5.35$.

By arguments similar to those already given for the pure birth process, we may now find that the probability $p_N(t)$ that at time t the population is of size N given that its initial size was i and $i > N$. It will be found that

$$p_N(t) = \binom{i}{N} e^{-\mu it} \left(e^{\mu t} - 1\right)^{i-N} \tag{8.5}$$

which is Eq. (1.4) in Pielou (1969). We may also find the expectation of the time to extinction. It follows from the modeling applied that the ultimate extinction is certain even though the birth and death rates are equal. Only if the population enjoys a positive rate of increase, is there a probability that the population will persist indefinitely. To find the probability of ultimate extinction, we must allow time t to tend to infinity. The reader is referred to Pielou (1969) book for details, and the stochastic generalization of the deterministic conclusions.

8.2.2 Growth of logistic population

Due to a shortage of resources, a stage is reached when the demands made by the existing population preclude further growth and the population is then at the "saturation level." This level is a value determined by the "carrying capacity" of the environment. The probability distribution of the population reaches then the stochastic equilibrium. At the end of Chapter 2, an Eq. (2.6) was derived for $d\ln N/dt$ implying a rapid decrease in individual growth rate with a population size N while N is small and less rapid as N becomes larger (confirmed by other findings).

8.2.3 Growth with age-dependent rates of birth and death I: The discrete time

Previous assumptions are now reversed, i.e., it is assumed that an individual's chances of reproducing and dying are functions of age, but that these chances are unaffected by the size of population in which density dependence begins to exert any influence. The model considered here is deterministic. It simplifies discussion to consider only the females in a bisexual population. The same arguments would also apply if one counts members of both sexes.

8.2.4 Growth with age-dependent rates of birth and death II: The continuous time

The author defines and then investigates new rates in terms of the instantaneous birth rate $b=B_t/N_t$, where B_t is the number of births occurring per unit of time at time t in the whole population of size N_t. It should be noted that B_t is a rate which is necessarily defined in terms of a unit of time and which varies continuously. The instantaneous death rate is consistently defined. Finally, life tables are obtained and their examples are given.

8.2.5 Growth of populations of two competing species

The earlier discussion is extended by considering populations of two species living together and competing with each other for the same limiting resource. Each population is inhibited not only by members of its own species but also by those of the other. New dynamical equations are proposed to describe this case.

8.2.6 Dynamics of host-parasite populations

In the preceding chapter, owing to competition, each species inhibited the multiplication of the other. An entirely different form of interaction between two species occurs when one species is a parasite (or predator) and the other is host (or prey). The two possibilities "parasite-host" and "predator-prey" are mathematically equivalent. However, as the parasite population grows, the number of hosts destroyed by parasites increases. Thus, the simple pair of equations describing the possible population growth is that originally described by Lotka (1920, 1922). However, the chapter also offers the derivation of extra equations that generalize the classical Lotka-Volterra model.

8.3 Spatial patterns in one-species populations

8.3.1 Spatial patterns represented by discrete distributions

When the population of sessile (or, at least, sedentary) organisms occupying a large area or volume is considered, it is clear that unless all the organisms are evenly spaced out, the density effect must vary from place to place. For organisms grouped in dense clusters, their mutual interference must be greater than at sites in which they are sparse. In this case, it is unreasonable to suppose that the birth and death rates per individual are functions of the number of members in the population as a whole. Consequences of this situation are discussed in the chapter.

8.3.2 The measurement of aggregation

We may need to measure the degree of aggregation (clumping, clustering, or contagion) of a population's spatial pattern. It would then be possible to compare the aggregation exhibited by a single species at different times or in different places, or to compare the aggregation found at a single place and time in populations of two different species. All such observations are of obvious ecological interest. Usually, two compared populations will differ in the mean density as well as in the degree of aggregation.

8.3.3 The patterns of individuals in a continuum

Here, one turns to the case in which an extended continuum, either an area or a volume is available to the organisms, and they may be found anywhere throughout it. Examples are individual plants scattered over an area of ground or micro-arthropods dispersed through a volume or soil. There are now no natural sampling units such as the discrete habitable units afforded, so the sampling units have to be arbitrarily defined. The commonest method of investigation, the pattern of a related population, is to sample it with randomly placed quadrats. They are small sample areas, which are usually, but not necessarily, square.

8.3.4 A pattern studied by discrete sampling

Quadrat sampling is only one of the methods that may be used to study the spatial pattern of a population of organisms dispersed over a continuous surface. The method suffers from the disadvantage that quadrats are not natural sampling units but necessarily are arbitrary. There is, however, a wholly different method of investigating the pattern of points in a plane. This is the so-called "plotless sampling." What is examined is the spacing of the individuals. There are two ways to proceed described in the chapter.

8.3.5 Patterns resulting from diffusion

The preceding discussion of spatial patterns in the chapters treated these patterns as static. However, it is clear that any pattern must have a history. To reach the locations in which one observes patterns, the organisms of a population must have moved, either actively or passively. The movements of animal populations may be two types: migration and diffusion. Examples of migration are provided by many bird species. In contrast to migration, diffusion consists in the apparently aimless, undirected movements of animals that seem to be wholly random. The mathematical analyses of the random walk and diffusion are further performed in one and two dimensions. The author includes the passage to the Fokker-Planck equation from the random walk (El-Wakil and Zahran, 2000; El-Wakil et al., 2001).

8.3.6 Patterns of ecological maps. Two-phase mosaics

In this chapter, the author considers properties of ecological patterns from the static viewpoint and discusses a (third) pattern of the three existing types. These patterns are exhibited by vegetatively reproducing plants, which commonly occur as extensive clumps of shoots. Regardless of whether the individual shoots of a clump are densely packed or well separated, it is reasonable to treat the clumps rather than the shoots as the entities whose pattern is to be studied. Considered are practical problems of sampling mosaics as well as non-random and anisotropic mosaics. A figure provides a random-set mosaic or S-mosaic.

8.4 Spatial relations of two or more species

8.4.1 Association between pairs of species

Individuals in discrete habitable units

Examined is a spatial pattern exhibited by a single species within limited area. However, the factors controlling and determining pattern are likely to many species rather than just one, and much may be learned by investigating the way in which species are associated with one another. If two co-occurring species are affected by the same environmental factors or if they exhibit an effect, favorable or not, on each other, their patterns will not be independent; the species will be associated, either positively or negatively. Association or the lack of it among pairs and groups of species is therefore of obvious ecological interest. As in the study of pattern in one-species populations, it is desirable to consider separately those species that occupy discrete habitable units, and those that may occur anywhere throughout an extended space of continuum (e.g., plankton organisms in a volume of water and plants in a meadow). Considerations are confined to organisms in discrete units and begin with the case of a single pair of species. Stability problems may be important in ecological populations (Upadhyay et al., 2000), the problems which are too scarcely discussed in the book. Also, thermodynamic aspects of complexity in ecosystems, Schneider and Kay (1994), are omitted.

8.4.2 Association between pairs of species

Individuals in a continuum

Studies of the association between species which occupy discrete units usually take no account of the spatial arrangement of the units. The mathematical methods for judging the association between, say, a pair of parasite species infecting a population of mammals are formally identical with those for testing the association between, for example, the eye colors of parents and children. The spatial arrangement of the sample units is disregarded except in so far the population studied is defined in terms of the geographical area it occupies. In testing for association between two plant species, it is customary to treat each quadrat as if it were a discrete sample unit and so use the same methods as those described in the previous chapter.

8.4.3 Segregation between two species

It was already shown in the previous chapter that when the association between two species of plants was examined, the results will be strongly influenced by both the spacing of the quadrats and their sizes. This is because what is being investigated is not so much interspecies relationships per se but rather joint, two-species patterns. Most of other factors affect these joint patterns. Thus, it is worthwhile to study the patterns of each species in relations to the other without regard to the pattern of either in relations to the ground. In the pursued investigation, discrete plants are assumed.

8.5 Many species populations

8.5.1 Species-abundance relations

Most ecological communities contain many species of organisms, so the species may vary greatly in their abundance from very common to very rare. Therefore, as soon as one attempts to study whole communities rather than the interrelations among a few chosen species, the question arises: How are the abundances of the different species distributed? There are also associated questions, formulated therein.

8.5.2 Ecological diversity and its measurement

When the species-abundance frequencies in an actual collection are well fitted by one or another of the theoretical distributions, the parameters of the fitted distribution are obviously suitable as descriptive statistics. If the distribution is the log-normal, the appropriate statistics are the estimates of S^n, the total number of species in the population, and σ^2, the variance of the logarithmic curve. If the distribution is negative binomial with k different than zero, the appropriate statistics are the estimates of S^n and k (a parameter I^n depends on sample size). The logarithmic series distribution has two parameters, α and X, but, since X is a function of sample size and S^n is assumed to the infinitely large, this leaves only α as a statistics that describes an intrinsic property of the population being sampled. The broken-stick distribution also has only a single descriptive parameter, S, the number of species (assumed to be the same in both sample and population).

8.5.3 The classification of communities

In ecology, most of the work on classification has been done by students of vegetation and such classification problem is considered here. The sample units are quadrats or larger stands of vegetation. When arbitrarily delimited quadrats are used, there is always a risk that the classification obtained may be markedly affected by quadrat size; if stands of vegetation with natural boundaries form the sample units, they may differ greatly in area. A major problem found by the classifier is whether classification is even appropriate.

8.5.4 The ordination of continuously varying communities

Quite often, the points representing the quadrats are diffusely scattered and any classification procedure is largely arbitrary. A way out of this difficulty is to ordinate the quadrats rather than to classify them. The purpose, as in classifications, is still to simplify and condense the mass of raw data yielded by vegetation sampling in the hope that relationships among the plant species and between them and the environmental variables will be manifested. Ordination consists in plotting n points in a space of fewer than s dimensions in such a way that none of the important features of the original s-dimensional pattern is lost. As a method of the results summarizing, ordination has great advantages over classification.

8.5.5 Canonical variate analysis

The previous chapter dealt with the way of simplifying and condensing the raw data obtained from a vegetation survey. Each quadrat, or other sample unit, yielded a vector of variate values (the amounts of the several species), and it was shown how, by the technique of principal component analysis, one can find a few linear combination of the variates that account for nearly all the variables in the data. Yet, the observations on the quantities of the different species in a unit of vegetation may constitute only one of the sets of observations made by an ecologist. These sets are described in the considered book along with the useful tools of analysis.

8.6 Feldman's (1969) review

Book Review: *Into the Ecology Breach An Introduction to Mathematical Ecology*. E. C. PIELOU. Wiley-Interscience, New York, 1969. x + 294 pp., illus. $14.95.

Ecologists at the present time appear to be divided into two groups. On the one hand, there are those who believe that ecological phenomena can be abstracted and modeled in such a way that mathematics can be brought to bear in their analysis. "Mathematics" here is taken to include numerical analysis through electronic computation. On the other hand, there are those who believe that the complexity of the phenomena is so great that no mathematical model can be successful and those who ignore all biological research which contains any mathematics. As a result of this polarization, the latter group has ceased to be able to evaluate the relevance or importance of the work of the former. With this valuable and timely book, Pielou has attempted to provide a basis for the resumption of communication between the two groups. It is probably the first text on mathematical ecology directed to the biologist with minimal mathematical background who wishes to learn what model analysis can accomplish and what some of the open problems are.

The book is conveniently divided into four quite self-contained parts in such a way that it is not necessary to proceed in the order they are presented in. However, most readers will prefer to attempt part 1, on population dynamics, first, since this is one of the oldest and best-known parts of the subject. Although it is brief, the treatment given should provide the conceptual base for the understanding of much of the recent literature centered around the competition equations.

Parts 2 and 3 are devoted to the spatial patterns in one- and multi-species populations, respectively, subjects to which Pielou herself has made significant contributions. Both discrete and continuous populations are treated, and the consequences of various sampling procedures, such as quadrat sampling and distance sampling, are investigated. In all cases, assumptions are clearly stated and the limitations of the resulting formulas adequately explained.

Whereas parts 2 and 3 seem to focus on plant ecology, the final part of the book is of wider interest. It contains a good discussion of the mathematics behind species-abundance relations, currently one of the hottest topics in statistical ecology. One of the characteristics of some recent work in mathematical ecology has been its lack of rigor. Precision has often been sacrificed for generality and realism. The danger in this procedure, of course, is that the conclusions of the subsequent "analysis," which are often qualitative conclusions, are then suspect. With this book, Pielou has attempted, on the whole successfully, to introduce the student to the more rigorous approach.

At times, there are anomalies in the degree of sophistication expected of the reader. For example, in the chapter "Patterns resulting from diffusion" the author includes the passage to the Fokker-Planck equation from the random walk. But she has chosen to omit the simple and elegant probability-generating function approach to the birth and death processes of Chapter 1 in favor of a more messy induction approach.

Again, in Chapter 5 when the competition equations are dealt with the local stability criterion for ordinary, differentia equations should have been invoked so that the reader could easily move relevant biological situations and data pertaining to the various models are omitted. One may therefore predict that the reader, if a biologist, will find his interest flagging as he proceeds through the calculations without seeing how they tie in with observation. On the other hand, sufficient references are given so that the enthusiastic and diligent reader can correct this deficiency for himself. Pielou has been successful in conveying to the reader the rationale behind the various statistical definitions and in showing where they are and are not satisfactory.

Areas where further research is needed are discussed, and often, the direction this research might most fruitfully take is outlined. Mathematical ecology has become a fashionable discipline. This volume should enable graduate students and researchers in ecology to evaluate the work in an area which to some has seemed mysterious and which others have taken on faith.

Marcus W. Feldman. Department of Biological Sciences, Stanford University, Stanford, California, USA.

References

El-Wakil, S.A., Elhandbay, A., Zahran, M.A., 2001. Fractional (space–time) Fokker–Planck equation. Chaos Solitons Fractals 12, 1035–1040.

El-Wakil, S.A., Zahran, M.A., 2000. Fractional Fokker–Planck equation. Chaos Solitons Fractals 12, 791–798.

Feldman, M.W., 1969. Into the Ecology Breach An Introduction to Mathematical Ecology E.C. Pielou. Wiley-Interscience, New York (Review of the book).

Lotka, A.J., 1920. Undamped oscillations derived from the law of mass action. J. Am. Chem. Soc. 42, 1595–1599.

Lotka, A.J., 1922. Contribution to the energetics of evolution. Proc. Natl. Acad. Sci. U. S. A. 8, 147–151.

Pielou, E.C., 1969. An Introduction to Mathematical Ecology. Wiley-Interscience, New York.

Schneider, E., Kay, J., 1994. Complexity and thermodynamics: towards a new ecology. Futures 24, 626–647.

Ulanowicz, R.E., 1997. Ecology, the Ascendent Perspective. Columbia University Press, New York.

Upadhyay, R.K., Iyengar, S.R.K., Rai, V., 2000. Stability and complexity in ecological systems. Chaos, Solitons Fractals 11, 533–542.

Further reading

Gaveau, B., Moreau, M., Toth, J., 1999a. Variational nonequilibrium thermodynamics of reaction-diffusion systems, I: the information potential. J. Chem. Phys. 111 (17), 7736–7747.

Gaveau, B., Moreau, M., Toth, J., 1999b. Variational nonequilibrium thermodynamics of reaction-diffusion systems, II: path integrals, large fluctuations, and rate constants. J. Chem. Phys. 111 (17), 7748–7757.

Gaveau, B., Moreau, M., Toth, J., 2001. Variational nonequilibrium thermodynamics of reaction-diffusion systems, III: progress variables and dissipation of energy and information. J. Chem. Phys. 115 (2), 680–690.

Goldstein, S., 1953. On diffusion by discontinuous movements and on the telegraph equation. Q. J. Mech. Appl. Math. 6, 290–312.

CHAPTER 9

Optimizing in ecological systems

9.1 Introducing the standard form of continuous optimization problem

The goal of this chapter is to derive classical necessary and locally sufficient optimality conditions for the minimum or maximum of functionals and then formulate and analyze the resulting theory of dynamic optimization.

To begin with, optimal continuous processes are recalled. They are described by a system of ordinary differential equations

$$\frac{dx_i}{dt} = f_i(x_1, x_2, \ldots x_s, t, u_1, u_2, \ldots\ldots u_r) \quad i = 1,2,\ldots,s \quad t_p \le t \le t_k \tag{9.1}$$

or in the vector form

$$\mathrm{d}\mathbf{x}/\mathrm{d}t = \mathbf{f}(\mathbf{x}, t, \mathbf{u}) \;\; t_p \le t \le t_k \tag{9.2}$$

where: $\mathbf{x}(t)$ – s-dimensional continuous vector function representing the state of the process in time t; $\mathbf{u}(t)$ – r-dimensional piecewise continuous vector function describing controls (decisions) in time t. The variable t may represent chronological time in a batch process or residence time in a stationary continuous process. This variable may also represent only a measure of time, like, e.g., a length variable which describes the distance from the inlet of a tubular reactor with perfect radial mixing.

Applications of model (9.1) for batch processes usually comprise various apparatuses with an ideal mixing. Yet, in the case of stationary one-dimensional continuous processes, applications involve systems in which basic state changes occur only in one well-defined direction characterized either by length or by time.

Control variables are often constrained. When constraints involve state variables, controls and time, they have the form of a system of inequalities

$$\psi_j(x_1, x_2, \ldots x_s, t, u_1, u_2, \ldots\ldots u_r) \le 0 \quad j = 1,2,\ldots,m \tag{9.3}$$

However, let us consider the simplest constraints involving only control variables, whose vector form is

$$\psi_j(u_1, u_2, \ldots\ldots u_r) \equiv \psi_j(\mathbf{u}) \le 0 \quad j = 1,2,\ldots,m \tag{9.4}$$

Complexity and Complex Ecological Systems. https://doi.org/10.1016/B978-0-443-19237-1.00002-8

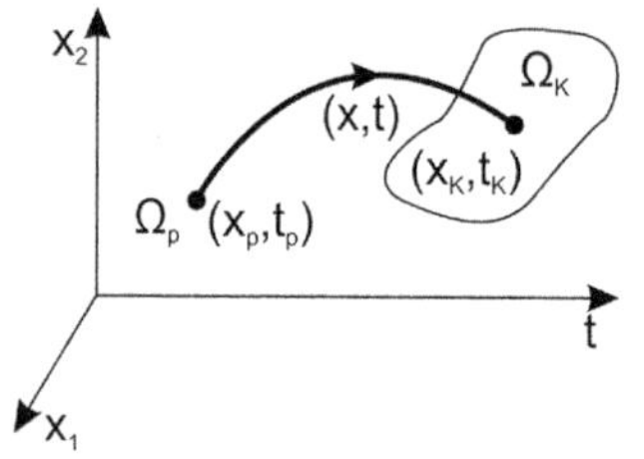

Fig. 9.1
Target trajectory arrives at final multitude Ω_k for an admissible control $\mathbf{u}(t) \in \mathbf{U}$.

In more general terms, constraints of this class are written as

$$\mathbf{u}(t) \in \mathbf{U} \tag{9.5}$$

where $\mathbf{U}$ is the regime of admissible values of $\mathbf{u}(t)$ in the r-dimensional space of controls. The control satisfying constraint (9.5) is called *admissible*. In addition, it is called *target control if* it transfers process state to a definite final manifold (also called final multitude) in the state space, Fig. 9.1, see the discussion of Eqs. (9.13), (9.14) below. The final manifold is defined mathematically and, in particular, it may be a straight line if prescribed are all final coordinates while the final time t_k is free, or it may be a point if both the final coordinates and the final time t_k are prescribed. In general, the final manifold is represented by a certain hypersurface Ω_k of a definite dimensionality.

Of course, in some problems, initial states may also be chosen optimally on a certain initial manifold (initial multitude).

Whenever time t appears explicitly in an optimization model (the case of so-called non-autonomous systems), then a geometric interpretation of basic components of the model is usually done in an extended (enlarged) state space which includes time t into the set of coordinates. It is then convenient to employ an extended state space of coordinates $x_1, x_2, \ldots .x_s$, x_{s+1} associated with the extended state vector

$$\tilde{\mathbf{x}} \equiv [x_1, x_2, \ldots .x_s, x_{s+1}]^{\mathrm{T}}$$

in which $x_{s+1} = t$. The coordinate x_{s+1} satisfies the trivial differential equation

$$\frac{dx_{s+1}}{dt} = f_{s+1} \equiv 1 \tag{9.6}$$

The vector notation of Eqs. (9.1), (9.2) assumes the form

$$\frac{d\tilde{\mathbf{x}}}{dt} = \mathbf{f}(\tilde{\mathbf{x}}, \mathbf{u}) \tag{9.7}$$

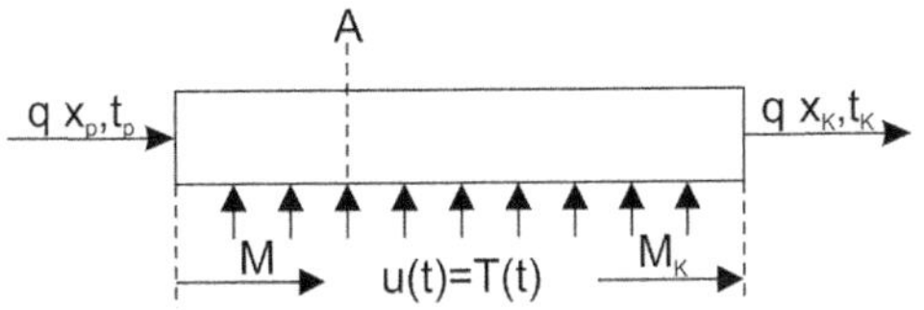

Fig. 9.2
Continuous decision process in a tubular reactor.

For a definite vector control $\mathbf{u}(t)$, the process in the extended space is interpreted geometrically as a motion along a curve called the trajectory, Fig. 9.1. The trajectory starts at initial point (t_p, x_p) residing on an initial multitude Ω_p and terminates at final point (t_k, x_k) residing on a final multitude Ω_k. In Fig. 9.1, the initial multitude is a point, whereas the final multitude is a two-dimensional surface.

It should be noted that any admissible control (i.e., control satisfying the condition $\mathbf{u}(t)\in\mathbf{U}$) will not always be capable of transferring process trajectories onto the final multitude Ω_k; it may so happen that trajectories always omit the multitude Ω_k. This may be particularly frequent when the dimension of the multitude Ω_k is small, i.e., when Ω_k is a point (zero-dimensional multitude). In this case, the formulation of the target control problem is impossible.

In order to make the quality index (optimization criterion) applicable to a relatively large number of practical cases, consider a steady control process in the tubular reactor schematized in Fig. 9.2. For concreteness, an isothermal chemical transformation with s independent chemical reactions is assumed. In this case, the state coordinates are concentrations of key components $x_1, x_2, \ldots, x_s$. The concentrations of remaining components $x_{s+1}, \ldots x_{s+2}, \ldots x_v$ can be expressed by key concentrations with the help of chemical invariants stemming from mass balances of chemical elements (Gadewar et al., 2001).

Let us assume a given mass flux of the reagent's stream, q(kg/h), and a known relationship linking economic prices per unit mass of the solution with concentrations $\tilde{C}(x_1, x_2, .x_s.., x_v)$. Concentrations $x_{s+1}, \ldots x_{s+2}, \ldots x_v$ are eliminated to find the function $C(x_1,x_2,..,x_s)$ which describes the unit price of the reacting stream in terms of coordinates of the state vector $\mathbf{x}$. The gross profit, defined as the difference between the value flux of outgoing stream and the value flux of incoming stream, is expressed by the simple formula

$$D = q\left(C\left(\mathbf{x}^k\right) - C\left(\mathbf{x}^p\right)\right) \tag{9.8}$$

However, in order to obtain a more appropriate optimization criterion, the reactor cooling cost (operational component), associated with the consumption of cold water, should be subtracted from the profit D. Let us designate by the symbol $c_m(T)$ the temperature-dependent function describing the unit mass price of the cooling medium. Designate by M the integral mass flux of the cooling medium calculated for the reactor part between the inlet cross-section and the considered cross-section, A. With these data, the reactor cooling cost can be determined as an integral

$$K_m = \int_{M(t_p=0)}^{M(t_k)} c_m(T)dM \tag{9.9}$$

Moreover, the capital cost K_i, associated with the equipment purchase, should be subtracted from the profit D. For tubular reactors of fixed diameter, cost K_i is proportional to the apparatus length, and, for the constant flow of the cooling medium per unit length of the reactor, to the final value of cooling mass flux, M^{k}. Therefore,

$$K_i = \lambda M^{\mathrm{k}} \tag{9.10}$$

where λ is a proportionality coefficient. Decrease in value of the equipment due to obsolescence or use may influence the value of λ which is then a decreasing function of time t.

Consequently, the optimization criterion (quality index) describing the process profit per unit mass flow of reacting stream q takes the form

$$S = \frac{D - K_m - K_i}{q} = -\int_0^{\tau^k} c_m(T)d\tau + C(\mathbf{x}^k) - C(\mathbf{x}^p) - \lambda\tau^k \tag{9.11}$$

($\tau = M/q$). The quantity $\tau = M/q$ is a certain dimensionless measure of residence time of the reacting mixture in the reactor.

It may be observed that the quality index S is the sum of a certain integral and a function dependent on state $\mathbf{x}$ and time t at the beginning and at the end of the process. Criteria of this structure appear very frequently in optimization; they are so-called Bolza functionals.

A general form of Bolza functionals to be investigated in further considerations is

$$S[\mathbf{x}(t), \mathbf{u}(t)] = \int_{t^p}^{t^k} f_0(\mathbf{x}(t), \mathbf{u}(t), t)dt + G(\mathbf{x}^k, t^k) - G(\mathbf{x}^p, t^p) \tag{9.12}$$

The optimization criterion (9.12) contains two scalar functions, f_0 and G.

The general optimization problem associated with the criterion (9.12) is to find such an admissible target control $\widehat{\mathbf{u}}(t)$ which renders the criterion (9.12) the maximum (minimum). Finding of optimal control $\widehat{\mathbf{u}}(t)$ is usually associated with simultaneous finding of some coordinates of extended initial state $\tilde{\mathbf{x}}(t^p)$ and extended final state $\tilde{\mathbf{x}}(t^k)$, and also the optimal trajectory $\widehat{\mathbf{x}}(t)$. Note that this task may also include finding an optimal final time t^{k} or an optimal duration t^{k} - t^{p}.

The necessary optimality conditions for functional (9.12) are investigated below for state constraints (9.2) and control constraints $\mathbf{u}(t) \in \mathbf{U}$, Eq. (9.5), as well as constraints imposed by some initial and final conditions for state coordinates and time. The latter are exemplified by the following boundary conditions (initial and final)

$$g_{pr}(t^p, \mathbf{x}^p) = 0 \quad r = 1, 2, \ldots, a \quad (a \text{ equations}) \tag{9.13}$$

$$g_{kr}\left(t^k, \mathbf{x}^k\right) = 0 \quad r = a+1, a+2 \ldots, a+\beta \quad (\beta \text{ equations}) \tag{9.14}$$

Eqs. (9.13), (9.14) require that in the extended state space the initial point of the trajectory resides on a certain manifold (hypersurface) Ω^{p}, and the final point of the trajectory—on the manifold (hypersurface) Ω^{k}, Fig. 9.1. Conditions (9.13) and (9.14) can therefore be written in the form

$$\left(\mathbf{x}(t^p), t^p\right) \in \Omega^p \tag{9.15}$$

$$\left(\mathbf{x}\left(t^k\right), t^k\right) \in \Omega^k \tag{9.16}$$

The simplest form of these conditions prescribes all coordinates of extended initial and final state, i.e., requires that $\mathbf{x}(t^{\text{p}}) = \overline{\mathbf{x}}_p$, $t^{\text{p}} = \overline{t}_p$ as well as $\mathbf{x}(t^{\text{k}}) = \overline{\mathbf{x}}_k$, $t^{\text{k}} = \overline{t}_k$. Other cases of conditions (9.15) and (9.16) are discussed in Section 9.2.2.

9.2 Dynamic programming investigation of optimal quality function

9.2.1 Hamilton-Jacobi-Bellman equation

Investigation of the performance criterion (9.12) is preceded by achieving an equivalent integral form associated with the so-called Lagrange structure. By defining the function

$$L = f_0(\mathbf{x}(t), \mathbf{u}(t), t) + \frac{dG}{dt} = f_0(\mathbf{x}(t), \mathbf{u}(t), t) + \frac{\partial G}{\partial t} + \sum_{i=1}^{s} \frac{\partial G}{\partial x_i} f_i \tag{9.17}$$

the problem is broken down to the maximizing performance function of the form

$$S[\mathbf{x}(t), \mathbf{u}(t)] = \int_{t^p}^{t^k} L_0(\mathbf{x}(t), \mathbf{u}(t), t)dt \tag{9.18}$$

The dynamic programming method (DP) is used to investigate the optimality conditions and optimization solution for the above criterion. For this purpose, we define the following optimal performance function (backward DP algorithm)

$$Q\left(\mathbf{x}^0, t^0\right) = max\, S[\mathbf{x}(t), \mathbf{u}(t)] \tag{9.19}$$

which describes the maximum performance index. Function $Q(\mathbf{x}^0, t^0)$ is obtained by moving along an optimal trajectory, which starts from an arbitrary point $(\mathbf{x}^0, t^0)$ and achieves the final manifold Ω^{k} in time t^{k}, prescribed or not. It is assumed that the corresponding control is

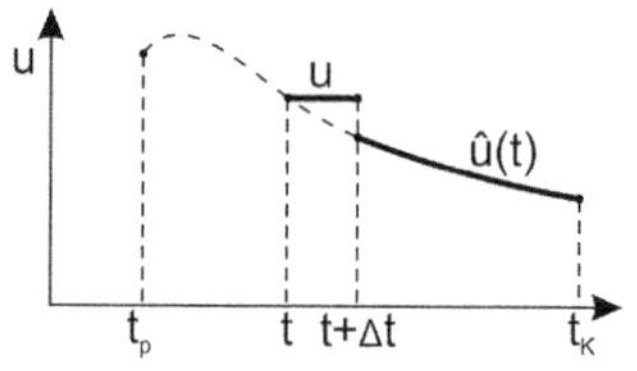

Fig. 9.3
Application of the optimality principle: When the backward DP algorithm is used in optimization, constant control approximation within a short-time period $(t, t+\Delta t)$ precedes an optimal control segment for the final period of the process $(t+\Delta t, t^k)$.

admissible, i.e., satisfies constraints (9.3). Since the point $(\mathbf{x}^0, t^0)$ is arbitrary, superscript zero is omitted in further considerations, i.e., $Q(\mathbf{x}^0, t^0)$.

$\equiv Q(\mathbf{x}, t)$.

Let us employ properties of Bellman's optimality principle (Aris, 1964; Bellman, 1967) incorporated into the backward DP algorithm. To derive a locally sufficient optimality condition under the assumption of the differentiability of function $Q(\mathbf{x}, t)$, we consider the summary effect resulting from optimal control applied during the final period of time $(t+\Delta t, t^k)$, and from constant non-optimal control applied during a short period $(t, t+\Delta t)$. Both periods are shown in Fig. 9.3, which also shows an approximation of constant control for a short-time period $(t, t+\Delta t)$ and a finite segment of optimal control for a finite final period $(t+\Delta t, t^k)$.

The performance index for the whole considered time period (t, t^k) is

$$S = Q(\mathbf{x} + \Delta\mathbf{x}, t + \Delta t) + L(\mathbf{x}(t), \mathbf{u}(t), t)\Delta t + 0(\varepsilon^2) \tag{9.20}$$

where $0(\varepsilon^2)$ represents second and higher order terms which vanish when Δt approaches zero. Retaining only linear terms in the Taylor series expansion for $Q(\mathbf{x}+\Delta\mathbf{x}, t+\Delta t)$, we obtain from Eq. (9.20)

$$S = Q(\mathbf{x}, t) + \sum_{i=1}^{s} \frac{\partial Q(\mathbf{x}, t)}{\partial x_i}\Delta x_i + \frac{\partial Q(\mathbf{x}, t)}{\partial t}\Delta t + L(\mathbf{x}(t), \mathbf{u}(t), t)\Delta t + 0(\varepsilon^2) \tag{9.21}$$

Since changes Δx_i and Δt are linked in state Eq. (9.1), then for small Δt and neglected terms $0(\varepsilon^2)$

$$\Delta x_i = f_i(\mathbf{x}, \mathbf{u}, t)\Delta t + 0(\varepsilon^2) \tag{9.22}$$

Therefore, after substituting Eq. (9.22) into Eq. (9.21)

$$S = Q(\mathbf{x}, t) + \sum_{i=1}^{s} \frac{\partial Q(\mathbf{x}, t)}{\partial x_i} f_i(\mathbf{x}, \mathbf{u}, t)\Delta t + \frac{\partial Q(\mathbf{x}, t)}{\partial t}\Delta t + L(\mathbf{x}(t), \mathbf{u}(t), t)\Delta t + 0(\varepsilon^2) \tag{9.23}$$

According to Bellman's optimality principle, (Bellman and Dreyfus, 1962; Aris, 1964), S has to be maximized in the whole period (t, t^{k}) by a suitable choice of control $\mathbf{u}$, which can be assumed constant for the small period Δt. Whence

$$\max S = \max_{\mathbf{u}} \left(\begin{array}{l} Q(\mathbf{x}, t) + \sum_{i=1}^{s} \frac{\partial Q(\mathbf{x}, t)}{\partial x_i} f_i(\mathbf{x}, \mathbf{u}, t)\Delta t \\ + \frac{\partial Q(\mathbf{x}, t)}{\partial t} \Delta t + L(\mathbf{x}, \mathbf{u}, t)\Delta t + 0(\varepsilon^2) \end{array} \right) \tag{9.24}$$

But, from the definition of the optimal performance function, Eq. (9.19), max $S = Q(\mathbf{x}, t)$. Consequently,

$$Q(\mathbf{x}, t) = \max_{\mathbf{u}} \left(\begin{array}{l} Q(\mathbf{x}, t) + \sum_{i=1}^{s} \frac{\partial Q(\mathbf{x}, t)}{\partial x_i} f_i(\mathbf{x}, \mathbf{u}, t)\Delta t \\ + \frac{\partial Q(\mathbf{x}, t)}{\partial t} \Delta t + L(\mathbf{x}, \mathbf{u}, t)\Delta t + 0(\varepsilon^2) \end{array} \right) \tag{9.25}$$

Since the maximizing operation is performed only with respect to controls $\mathbf{u}$, the terms not containing $\mathbf{u}$ can be moved out of the bracket in Eq. (9.25) and reduced. Dividing this simplified result by Δt yields an equation

$$\frac{\partial Q(\mathbf{x}, t)}{\partial t} = -\max_{\mathbf{u}} \left(\sum_{i=1}^{s} \frac{\partial Q(\mathbf{x}, t)}{\partial x_i} f_i(\mathbf{x}, \mathbf{u}, t) + L(\mathbf{x}, \mathbf{u}, t) + \frac{0(\varepsilon^2)}{\Delta t} \right) \tag{9.26}$$

Transforming Eq. (9.26) to the limit when Δt approaches zero yields the so-called Hamilton-Jacobi-Bellman equation (HJB equation)

$$\frac{\partial Q(\mathbf{x}, t)}{\partial t} + \max_{\mathbf{u}} \left(\sum_{i=1}^{s} \frac{\partial Q(\mathbf{x}, t)}{\partial x_i} f_i(\mathbf{x}, \mathbf{u}, t) + L(\mathbf{x}, \mathbf{u}, t) \right) = 0 \tag{9.27}$$

This equation pertains directly to the performance index (9.12) and Lagrange function L. In terms of function f_0 appearing in Eq. (9.12), the related HJB equation is obtained by substitution of Eq. (9.17) into Eq. (9.27), which yields

$$\frac{\partial R(\mathbf{x}, t)}{\partial t} + \max_{\mathbf{u}} \left(\sum_{i=1}^{s} \frac{\partial R(\mathbf{x}, t)}{\partial x_i} f_i(\mathbf{x}, \mathbf{u}, t) + f_0(\mathbf{x}, \mathbf{u}, t) \right) = 0 \tag{9.28}$$

where the bracketed expression will be called Hamiltonian expression. This HJB equation contains the optimal function $R(\mathbf{x}, t)$ defined as.

$$R(\mathbf{x}, t) = Q(\mathbf{x}, t) + G(\mathbf{x}, t) \tag{9.29}$$

where G is the function appearing in Bolza functional (9.12). Further on, we use the optimal function $R(\mathbf{x}, t)$, Eq. (9.28) rather than $Q(\mathbf{x}, t)$.

Eq. (9.28) is one of the most important results of the optimization theory. The feedback form of optimal control is determined from this equation in terms of state and time, $\mathbf{u}(\mathbf{x}, t)$. Whenever the maximum of bracket expression of (9.28) has a stationary nature, Eq. (9.28) is equivalent to the set of $r+1$ equations

$$\frac{\partial R(\mathbf{x}, t)}{\partial t} + \sum_{i=1}^{s} \frac{\partial R(\mathbf{x}, t)}{\partial x_i} f_i(\mathbf{x}, \mathbf{u}, t) + f_0(\mathbf{x}, \mathbf{u}, t) = 0 \tag{9.30}$$

$$\sum_{i=1}^{s} \frac{\partial R(\mathbf{x}, t)}{\partial x_i} \frac{\partial f_i(\mathbf{x}, \mathbf{u}, t)}{\partial u_l} + \frac{\partial f_0(\mathbf{x}, \mathbf{u}, t)}{\partial u_l} = 0 l = 1,\ldots r \tag{9.31}$$

Eq. (9.31) is obtained by differentiating the bracket expression of Eq. (9.28) with respect to all controls and setting the derivatives to zero, i.e., the equation results as the stationarity condition of Eq. (9.28).

The general solving procedure for Eq. (9.28) is as follows:

1. The bracketed Hamiltonian expression is maximized with respect to controls $u_1, u_2, \ldots u_r$, subject to constraint $\mathbf{u}(t) \in \mathbf{U}$ while keeping state variables x_i and partial derivatives $\partial R/\partial x_i$ constant. This is a typical problem of static optimization for the function of many variables u_i, If this problem is solved analytically, then the optimal control $\hat{\mathbf{u}}$ is obtained as a function $\hat{\mathbf{u}}(\partial R/\partial \mathbf{x}, \mathbf{x}, t)$.
2. Function $\hat{\mathbf{u}}(\partial R/\partial \mathbf{x}, \mathbf{x}, t)$is substituted into Eq. (9.8), which then assumes the form

$$\frac{\partial R(\mathbf{x}, t)}{\partial t} + \sum_{i=1}^{s} \frac{\partial R(\mathbf{x}, t)}{\partial x_i} f_i(\mathbf{x}, \hat{\mathbf{u}}(\partial R/\partial \mathbf{x}, \mathbf{x}, t), t) + f_0(\mathbf{x}, \hat{\mathbf{u}}(\partial R/\partial \mathbf{x}, \mathbf{x}, t), t) = 0 \tag{9.32}$$

   The above first-order partial differential equation is in general nonlinear. This equation should be solved with the boundary condition.

$$R(\mathbf{x}_k, t_k) = G(\mathbf{x}_k, t_k) \tag{9.33}$$

   which follows from the definition of functions R and Q in Eqs. (9.29), (9.19). Variables x_k and t_k must satisfy equations of final manifold Ω_k, Eq. (9.14). The result of solving Eq. (9.32) is the function $R(\mathbf{x}, t)$.
3. By substituting an analytical expression for $R(\mathbf{x}, t)$ into the function $\hat{\mathbf{u}}(\partial R/\partial \mathbf{x}, \mathbf{x}, t)$ in Eq. (9.32), optimal control is obtained in the form $\hat{\mathbf{u}} = \phi(\mathbf{x}, t)$.
4. The optimal initial enlarged state $(\mathbf{x}_p, t_p)$ is determined on the initial manifold Ω^{P} so that $Q(\mathbf{x}_p, t_p)$=maximum at $(\mathbf{x}_p, t_p) \in \Omega_p$. Note that the procedure involves maximizing $Q = R(\mathbf{x}, t) - G(\mathbf{x}, t)$ subject to the constraint (9.13).

5. Substitution of optimal function $\widehat{\mathbf{u}} = \phi(\mathbf{x}, t)$ into state Eq. (9.1) and integration of differential equations obtained starting at the optimal initial point $(\mathbf{x}_p, t_p) \in \Omega_p$ gives the optimal trajectory $\widehat{\mathbf{x}}(t)$. Moreover, substitution of $\widehat{\mathbf{x}}(t)$ into $\widehat{\mathbf{u}} = \phi(\mathbf{x}, t)$ leads to the simplest and usually required form of optimal control, described by function $\mathbf{u}(t)$.

9.2.2 Hamiltonian, adjoint equations, and canonical set

To proceed further, the following function is considered

$$B(\mathbf{x}, t, \mathbf{u}) \equiv \frac{\partial R(\mathbf{x}, t)}{\partial t} + \sum_{i=1}^{s} \frac{\partial R(\mathbf{x}, t)}{\partial x_i} f_i(\mathbf{x}, \mathbf{u}, t) + f_0(\mathbf{x}, \mathbf{u}, t) \tag{9.34}$$

As it follows from Eq. (9.28), function $B(\mathbf{x}, t, \mathbf{u})$ satisfies the conditions

$$B(\mathbf{x}, t, \mathbf{u}) \leq 0 \text{ for all points } (\mathbf{x}, t, \mathbf{u}) \tag{9.35}$$

and

$$B(\mathbf{x}(t), t, \mathbf{u}(t)) = 0 \text{ for arbitrary optimal process } (\mathbf{x}(t), \mathbf{u}(t)) \tag{9.36}$$

In other words

$$\max_{\mathbf{u} \in \mathbf{U}} B(\mathbf{x}, t, \mathbf{u}) = 0 \tag{9.37}$$

This formula means that for an arbitrary point $\tilde{\mathbf{x}} = (\mathbf{x}, t)$ a point $\mathbf{u}(\mathbf{x},t)$ in $\mathbf{U}$ can be found that satisfies $B(\mathbf{x},t,\mathbf{u})=0$. This property is depicted in Fig. 9.4, for a constant $\tilde{\mathbf{x}}$.

Assume that the function $R(\mathbf{x}, t)$ possesses second partial derivatives and the function $f_i(\mathbf{x}, t, \mathbf{u})$ possesses first partial derivatives. (These are fairly strict requirements which are not always satisfied.) Let us fix some control $\mathbf{u}$ in set $\mathbf{U}$. Then, it follows from Eqs. (9.35), (9.36) that function $B(\mathbf{x}, t, \mathbf{u})$ achieves the maximum with respect to x_i and t (Fig. 9.4 for $\mathbf{u}$=constant), which means that the following partial derivatives must vanish.

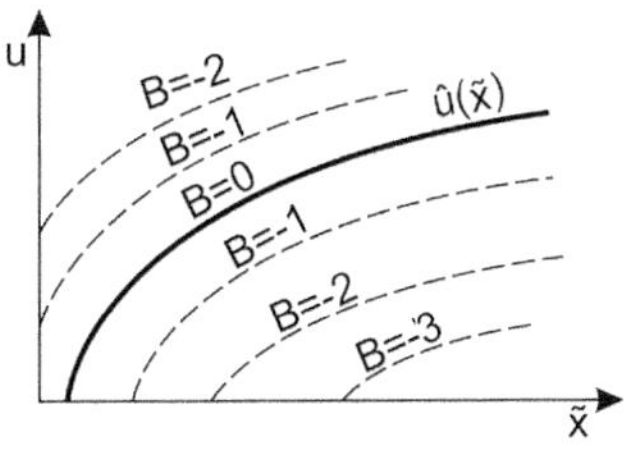

Fig. 9.4
Interpretation of properties of function $B(\mathbf{x}, t, \mathbf{u}) \equiv B(\tilde{\mathbf{x}}, \mathbf{u})$.

$$\frac{\partial B(\mathbf{x}, t, \mathbf{u})}{\partial x_i} = \sum_{i=1}^{s} \frac{\partial^2 R}{\partial x_i \partial x_j} f_i + \sum_{i=1}^{s} \frac{\partial R}{\partial x_i} \frac{\partial f_i}{\partial x_j} + \frac{\partial^2 R}{\partial t \partial x_j} + \frac{\partial f_0}{\partial x_j} = 0 \tag{9.38}$$

$$\frac{\partial B(\mathbf{x}, t, \mathbf{u})}{\partial t} = \sum_{i=1}^{s} \frac{\partial^2 R}{\partial x_i \partial t} f_i + \sum_{i=1}^{s} \frac{\partial R}{\partial x_i} \frac{\partial f_i}{\partial t} + \frac{\partial^2 R}{\partial t^2} + \frac{\partial f_0}{\partial t} = 0 \tag{9.39}$$

In order to transform Eqs. (9.38), (9.39), total derivatives of quantities $\partial R/\partial x_i$ and $\partial R/\partial t$ with respect to time are calculated. Since R depends solely on variables x_i and t, we find

$$\begin{aligned} \frac{d}{dt}\left(\frac{\partial R}{\partial x_i}\right) &= \sum_{i=1}^{s} \frac{\partial^2 R}{\partial x_i \partial x_j} \frac{dx_i}{dt} + \frac{\partial^2 R}{\partial x_j \partial t} \\ &= \sum_{i=1}^{s} \frac{\partial^2 R}{\partial x_i \partial x_j} f_i + \frac{\partial^2 R}{\partial t \partial x_j} \end{aligned} \tag{9.40}$$

and

$$\frac{d}{dt}\left(\frac{\partial R}{\partial x_i}\right) = \sum_{i=1}^{s} \frac{\partial^2 R}{\partial x_i \partial t} f_i + \frac{\partial^2 R}{\partial t^2} \tag{9.41}$$

Employing Eq. (9.38) in (9.39), (9.40), and (9.41) yields

$$\frac{d}{dt}\left(\frac{\partial R}{\partial x_j}\right) = -\sum_{i=1}^{s} \frac{\partial R}{\partial x_i} \frac{\partial f_i}{\partial x_j} - \frac{\partial f_0}{\partial x_j} \tag{9.42}$$

$$\frac{d}{dt}\left(\frac{\partial R}{\partial t}\right) = -\sum_{i=1}^{s} \frac{\partial R}{\partial x_i} \frac{\partial f_i}{\partial t} - \frac{\partial f_0}{\partial t} \tag{9.43}$$

Relationships (9.37), (9.42), and (9.43) can be written in a more lucid form after introducing the so-called adjoint variables (also called conjugate variables) z_i and z_t defined as:

$$\frac{\partial R(\mathbf{x}, t)}{\partial x_i} \equiv z_i \quad i = 1,\ldots,s \tag{9.44}$$

$$\frac{\partial R(\mathbf{x}, t)}{\partial t} \equiv z_t \tag{9.45}$$

and the Hamiltonian of the optimization problem

$$H(\mathbf{z}, \mathbf{x}, \mathbf{u}, t) = \sum_{i=1}^{s} z_i f_i + f_0 \tag{9.46}$$

Eqs. (9.37) and (9.42)–(9.46) then yield

$$z_t + \max_{\mathbf{u} \in \mathbf{U}} H(\mathbf{z}, \mathbf{x}, \mathbf{u}, t) = 0 \tag{9.47}$$

$$\frac{dz_i}{dt} = -\frac{\partial H}{\partial x_i}; i = 1,2,.,.. \tag{9.48}$$

$$\left(\frac{dz_t}{dt} = -\frac{\partial H}{\partial t}\right) \tag{9.49}$$

Eq. (9.49) is in brackets, since it follows from the remaining ones. However, as shall soon be seen, a *discrete* counterpart of Eq. (9.49) exists, which is a truly independent equation. This independence makes it possible to satisfy an extra optimality condition for an optimal interval of discrete time θ^n, (see further text).

It is easy to see that in terms of Hamiltonian (9.46), the state Eq. (9.1) can be written in the form

$$\frac{dx_i}{dt} = \frac{\partial H}{\partial z_i} \quad i = 1,2,..,s \quad t_p \leq t \leq t_k \tag{9.50}$$

Moreover, after calculating the derivative d*H*/d*t* for an optimal process and using Eq. (9.47), it is shown that Eq. (9.49) follows from the three remaining equations, and it does not have be used in calculations if other equations are employed. Eqs. (9.48)–(9.50) are called *canonical.*

Eqs. (9.47)–(9.50) constitute a set of necessary (and locally sufficient) optimality conditions. In terms of the enlarged state vector $\tilde{\mathbf{x}} \equiv (\mathbf{x}, t)$ and enlarged adjoint vector $\tilde{\mathbf{z}} \equiv (\mathbf{z}, z_t)$ as well as the enlarged Hamiltonian

$$\tilde{H}(\tilde{\mathbf{z}}, \tilde{\mathbf{x}}, \mathbf{u}) \equiv H(\mathbf{z}, \mathbf{x}, \mathbf{u}, t) + z_t \tag{9.51}$$

necessary optimality conditions can be written in an alternative form

$$\max_{\mathbf{u} \in \mathbf{U}} \tilde{H}(\tilde{\mathbf{z}}, \tilde{\mathbf{x}}, \mathbf{u}) = 0 \tag{9.52}$$

$$\frac{d\tilde{\mathbf{z}}}{dt} = -\frac{\partial \tilde{H}}{\partial \tilde{\mathbf{x}}} \tag{9.53}$$

$$\frac{d\tilde{\mathbf{x}}}{dt} = \frac{\partial \tilde{H}}{\partial \tilde{\mathbf{z}}} \tag{9.54}$$

Eqs. (9.47), (9.52) describe Pontryagin's maximum principle. In accordance with this principle an optimal control $\mathbf{u}(t)$ renders the Hamiltonian H or $\tilde{H}$a maximum at each time instant, and the optimum value of $\tilde{H}$equals zero. However, in order to solve the canonical equations, we need to supplement them by boundary conditions for both state and adjoint variables. This is the subject matter of the next section.

9.2.3 Transversality conditions

The necessary optimality conditions obtained above apply to every optimal trajectory regardless the boundary conditions (9.13) and (9.14). The effect of boundary conditions is manifested, however, in the form of transversality conditions, i.e., relations that must be satisfied due to variations of initial and final states. Only stationary variations are considered here.

If follows from the definition of optimal function $Q(\mathbf{x}, t)$ that the contribution of boundary conditions to the first variation of the functional S is:

$$\delta S = \left[\sum_{i=1}^{s} \frac{\partial Q}{\partial x_i} \delta x_i + \frac{\partial Q}{\partial t} \delta t \right]_{t=t_k, x=x_k}^{t=t_p, x=x_p} \tag{9.55}$$

This expression holds for changes calculated only along extremals, i.e., lines satisfying their canonical equations and equation of Hamiltonian maximum. Since Eqs. (9.29), (9.44), and (9.45) imply the equalities

$$\frac{\partial Q(\mathbf{x}, t)}{\partial x_i} = \frac{\partial R(\mathbf{x}, t)}{\partial x_i} - \frac{\partial G(\mathbf{x}, t)}{\partial x_i} = z_i - \frac{\partial G(\mathbf{x}, t)}{\partial x_i} \tag{9.56}$$

$$\frac{\partial Q(\mathbf{x}, t)}{\partial t} = \frac{\partial R(\mathbf{x}, t)}{\partial t} - \frac{\partial G(\mathbf{x}, t)}{\partial t} = z_i - \frac{\partial G(\mathbf{x}, t)}{\partial t} \tag{9.57}$$

then Eq. (9.55) may be transformed so that the variation δS is expressed by adjoint variables at the beginning and at the end of the trajectory. Substituting Eqs. (9.56), (9.57) in Eq. (9.55) and employing the property of vanishing variation in the optimal process, yields the following equations

$$\sum_{i=1}^{s} \left(z_i(t_p) - \frac{\partial G(\mathbf{x}_p, t_p)}{\partial x_{ip}} \right) \delta x_{ip} + \left(z_i(t_p) - \frac{\partial G(\mathbf{x}_p, t_p)}{\partial t_p} \right) \delta t_p = 0 \tag{9.58}$$

$$\sum_{i=1}^{s} \left(z_i(t_k) - \frac{\partial G(\mathbf{x}_k, t_k)}{\partial x_{ik}} \right) \delta x_{ik} + \left(z_i(t_k) - \frac{\partial G(\mathbf{x}_k, t_k)}{\partial t_k} \right) \delta t_k = 0 \tag{9.59}$$

These formulas must be satisfied for all permissible variations of enlarged initial state and for all permissible variations of enlarged final state. Consistently with Eqs. (9.13), (9.14), admissible variations ($\delta \mathbf{x}_p$, δt_p) and ($\delta \mathbf{x}_k$, δt_k) satisfy, in vector notation, the following conditions

$$\frac{\partial g_p(\mathbf{x}_p, t_p)}{\partial \mathbf{x}^p} \delta \mathbf{x}^p + \frac{\partial g_p(t^p, \mathbf{x}_p)}{\partial t^p} \delta t^p = 0 \quad \mathbf{g}_p = \left(g_{p1}, \ldots g_{p\alpha} \right) \tag{9.60}$$

$$\frac{\partial g_k(t^k, \mathbf{x}^k)}{\partial \mathbf{x}^k} \delta \mathbf{x}^k + \frac{\partial g_k(t^k, \mathbf{x}^k)}{\partial t^k} \delta t^k = 0 \quad \mathbf{g}_k = \left(g_{k1}, \ldots g_{k\beta} \right) \tag{9.61}$$

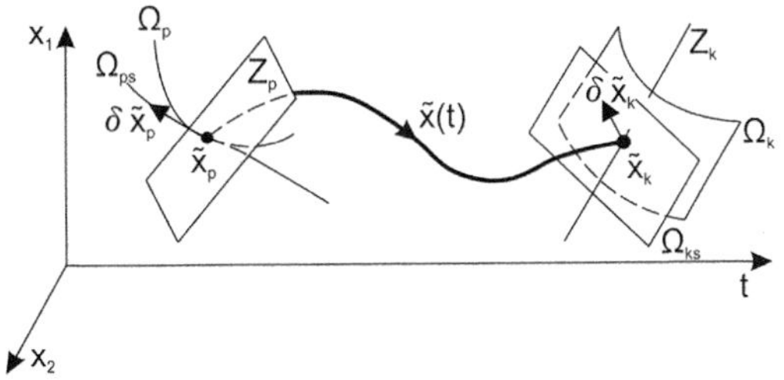

Fig. 9.5
Interpretation of transversality conditions in the case when $s=2$, $\alpha=2$, and $\beta=1$.

These conditions describe the requirement that the initial point of optimal trajectory must reside on the initial manifold Ω_p whereas the final point—on the final manifold Ω_k. Eqs. (9.60), (9.61) describe the tangency of the enlarged vector $(\delta\mathbf{x}_p, \delta t_p)$ to the initial manifold Ω_p and the tangency of enlarged vector $(\delta\mathbf{x}_k, \delta t_k)$ to the final manifold Ω_k. The set of such vectors constitutes linear manifolds Ω_{ps} and Ω_{ks} tangent to Ω_p and Ω_k, respectively, in points $[\mathbf{x}(t_p), t_p]$ and $[\mathbf{x}(t_k), t_k]$.

Eqs. (9.58)–(9.61) describe respectively the orthogonality of the enlarged vector

$$\left[\mathbf{z}(t_p) - \frac{\partial G(\mathbf{x}_p, t_p)}{\partial x_p}, z_t(t_p) - \frac{\partial G(\mathbf{x}_p, t_p)}{\partial t_p}\right] \tag{9.62}$$

to the initial manifold Ω_{ps}, and of the enlarged vector

$$\left[\mathbf{z}(t_k) - \frac{\partial G(\mathbf{x}_k, t_k)}{\partial \mathbf{x}_k}, z_t(t_k) - \frac{\partial G(\mathbf{x}_k, t_k)}{\partial t_k}\right] \tag{9.63}$$

to the final manifold Ω_{ks} (consider Fig. 9.5 for the case $s=2$). These orthogonality conditions are called *transversality conditions* of the optimization problem.

Summing up, α initial conditions (9.13) and β final conditions (9.14) dictate that the initial manifold Ω_p is $(s+1-\alpha)$ dimensional hypersurface in the spacetime $\mathbf{x}$, t. On the other hand, the final manifold is $(s+1-\beta)$ dimensional hypersurface in the spacetime $\mathbf{x}$, t. Tangent manifolds Ω_{ps} and Ω_{ks} have therefore dimensions $s+1-\alpha$ and $s+1-\beta$. Two vector sets (9.62) and (9.63), satisfying respectively Eqs. (9.58), (9.59), constitute linear manifolds Z_p and Z_k, which are respectively orthogonal to the manifolds Ω_{ps} and Ω_{ks}. The manifold Z_p is α-dimensional, and the manifold $Z_k-\beta$-dimensional.

As an example, consider the space of variables x_1, x_2, t, Fig. 9.5. In this case, $s=2$ and $s+1=3$.

Assume that the initial manifold is depicted by two Eq. (9.13) having the form $g_{1\text{p}}(t_p, x_{1p}, x_{2p})=0$ and $g_{2p}(t_p, x_{1p}, x_{2p})=0$. This means that $\alpha=2$, and that both initial manifolds Ω_p and Ω_{ps} have the dimension $s+1-a=1$. Therefore, Ω_p is a certain line (curve), whereas Ω_{ps} is a straight line. The dimension of the manifold $Z_p = \alpha = 2$. Since this manifold is a two-dimensional linear manifold, we deal with a straight line.

The final manifold Ω_k is described by single Eq. (9.14) of the form $g(t_k, x_{1k}, x_{2k}) = 0$. Therefore, $\beta = 1$ and manifolds Ω_k and Ω_{ks} have the dimension $s+1-\beta = 2$. Thus, Ω_k is a two-dimensional surface and Ω_{ks} is a plane. The dimension of the manifold Z_k equals $\beta = 1$; thus, it is a straight line.

Transversality conditions, i.e., relations resulting from the simultaneous satisfaction of Eqs. (9.58)–(9.61) supplement, respectively, the number of α initial conditions (9.13) to the complete number of $s+1$ initial conditions and the number of β final conditions (9.14) to the complete number of $s+1$ final conditions. As a result, we have at our disposal $2s+2$ boundary conditions for $2s+2$ differential Eqs. (9.53), (9.54).

Simple special cases of transversality conditions

The following important definition should be kept in mind:

Whenever an arbitrary variable x_{jp} or x_{jk} does not appear in equations of initial conditions (9.13) or final conditions (9.14), then this variable is called free or unspecified. An analogous terminology applies to times t_p or t_k.

Variations δx_{jp} or δx_{jk} of free variables, Eqs. (9.60), (9.61), may be of arbitrary sign and values. Consequently, Eq. (9.58) implies the following relationship

$$z_j(t_p) = \frac{\partial G(\mathbf{x}_p, t_p)}{\partial x_{jp}} \tag{9.64}$$

for a free variable x_{jp} at the beginning of the optimal trajectory. An analogous relationship

$$z_l(t_k) = \frac{\partial G(\mathbf{x}_k, t_k)}{\partial x_{lk}} \tag{9.65}$$

holds for a free variable x_{jp} at the end of the optimal trajectory.

Variations of free end times δt_p or δt_k may also be of arbitrary sign and value. In the case of a free final time, t_k, frequent in practice,

$$z_l(t_k) = \frac{\partial G(\mathbf{x}_k, t_k)}{\partial x_{lk}} \tag{9.66}$$

An analogous formula holds in the case of free initial time, t_p.

Eqs. (9.64)–(9.66) define boundary conditions for free coordinates of the enlarged state at the beginning or at the end of the optimal trajectory.

Lagrange forms of the performance functional

Whenever the functional (9.12) has the Lagrange form, i.e., when $G=0$, the boundary conditions for free state variables and time require the disappearance of corresponding adjoint variables. For example, in the case of a free state coordinate $x_l(t_k)$

$$z_t(t_k) = 0(G \equiv 0) \tag{9.67}$$

whereas in the case of a free final time t_k

$$z_l(t_k) = 0(G \equiv 0) \tag{9.68}$$

Eq. (9.67) along with Eq. (9.47) states that in an optimal process with free final time the final Hamiltonian vanishes, i.e.,

$$\tilde{H}(t_k) = 0 \tag{9.69}$$

In the method of variational calculus, momenta-involving counterparts of relationships (9.67)–(9.69) are known as *natural boundary conditions.*

If any coordinate of the initial or the final state, e.g., $x_r\ (t_k)$, is prescribed, then the admissible variation of this coordinate equals zero (the same conclusion pertains to the initial or final time). In this case, the corresponding coordinate of adjoint vector, $z_r\ (t_k)$, is free (undetermined). Consequently, this coordinate does not appear in the boundary conditions.

Whenever all coordinates of initial state and initial time are constant, then the following simple boundary conditions hold:

$$x_{ip} = \bar{x}_{ip} \quad i = 1, 2, \ldots, s \tag{9.70}$$

$$t_p = \bar{t}_p \quad i = 1, 2, \ldots, s \tag{9.71}$$

where $\bar{x}_{ip}$ and $\bar{t}_p$ are prescribed numerical values of x_{ip} and t_p. In the described case, the manifold of initial states, Ω_p, is reduced to a point in the enlarged state space ($\mathbf{x}$, t). If also the final point of the trajectory is fixed, i.e., $x_{ik} = \bar{x}_{ik}$ and $t_k = \bar{t}_k$, for $i=1, 2, \ldots s$ then the optimal trajectory starts at a fixed initial point of the spacetime ($\tilde{\mathbf{x}} = \mathbf{x}$, t) and ends at a fixed final point of this spacetime. Manifolds Ω_p and Ω_k are then points, and the boundary conditions do not contain any adjoint coordinates. These coordinates are undetermined (undefined) at the beginning and at the end of the optimal trajectory because the knowledge of $2s+2$ conditions for x_{ip}, x_{ik}, t_p, and t_k, is sufficient to solve the canonical set composed of state and adjoint equations. In the considered case, the form of function G in the performance index (9.12) does not have any influence on the solution of the optimization problem.

9.2.4 Two simple examples

Example 1. Find the maximum of the criterion

$$S = x(\bar{t}_k) - x(\bar{t}_p)$$

subject to:
the differential constraint

$$\frac{dx}{dt} = u - x,$$

boundary conditions

$$\bar{t}_p = 0, \quad \bar{t}_k = 1, \quad x(\bar{t}_p) = 0, \quad x(\bar{t}_k) = 1,$$

and the control constraint

$$-U \leq u \leq U.$$

where $U > 0$. The above model describes the problem of maximum temperature increase of a body heated by a medium having constrained temperature u.

Solution

The process Hamiltonian is

$$H = z(u - x)$$

and the condition of maximum H implies the optimal control

$$u = \begin{cases} -U \in z < 0 \\ \text{indefinite}, \ z = 0 \\ -U \in z > 0 \end{cases}$$

The adjoint equation has the form

$$\frac{dz}{dt} = -\frac{\partial H}{\partial x} = z$$

and the final condition for the adjoint variable z, Eq. (9.65), is

$$z(\bar{t}_k) = \frac{\partial G}{\partial x_k} = 1$$

(note that $G = x$). Integrating the adjoint equation yields

$$z(t) = z(\bar{t}_k) \exp(t - \bar{t}_k) = \exp(t - 1)$$

Therefore, $z(t) > 0$. This proves that the optimal control $\hat{\mathbf{u}} = U$, i.e., **u** attains the maximum admissible value.

Example 2. Find the minimum of the criterion

$$S = x(\bar{t}_k) - x(\bar{t}_p)$$

subject to:
the differential constraint

$$\frac{dx}{dt} = u - x,$$

boundary conditions

$$\bar{t}_p = 0, \ \ \bar{t}_k = 1, \ \ x(\bar{t}_p) = 0, \ \ x(\bar{t}_k) = 1,$$

and the control constraint

$$-U \leq u \leq U$$

where $U > 0$.

Solution

In this case, minimizing of S is required, which leads to the optimal solution $\hat{\mathbf{u}} = -U$, i.e., $\mathbf{u}$ attains the minimum admissible value.

The details regarding Hamiltonian behavior and a simple interpretation of the above conclusions for processes of body heating (cooling) by a heating (cooling) medium are left to the reader.

■

9.3 Continuous maximum principle

9.3.1 Basic algorithm and its modifications

Analytical difficulties associated with solving the Hamilton-Jacobi-Bellman equation are considerable; thus, this fundamental equation its seldom employed in optimization calculations. This equation constitutes, however, an important relationship for interpreting basic relations in the method of Pontryagin's Maximum Principle, which is one of the most efficient methods of dynamic optimization (Pontryagin et al., 1962; Halkin, 1966; Fan and Wang, 1964; Fan, 1966; Leitman, 1966, 1981; Boltyanski, 1971, 1973; Findeisen et al., 1977; Sieniutycz, 1991, and many others).

As shown in Section 9.2, under strong assumptions regarding differentiability of $Q(\mathbf{x}, t)$ and convexity of constraining sets, maximum principle can be derived from Bellman's principle of optimality in the form of the Hamilton-Jacobi-Bellman Eq. (9.28). However, maximum principle can also be derived in a way independent of Bellman's principle and HBJ equation. Such derivation (Pontryagin et al., 1962) uses much weaker assumptions than those necessary to derive Eq. (9.28).

Regardless the above subtleties, Pontryagin's maximum principle can be summarized in the form of following theorem:

Let $\mathbf{u}(t)$, $t_p \leq t \leq t_k$, be a piecewise continuous vector function describing an admissible, target control which ensures the state change from $(\mathbf{x}(t^p), t^p) \in \Omega^p$ to $(\mathbf{x}(t^k), t^k) \in \Omega^k$. Let $\mathbf{x}(t)$ is the process trajectory. Then, for the optimality of the process $\mathbf{u}(t)$, $\mathbf{x}(t)$, $t_p \leq t \leq t_k$, i.e., for the maximum of performance criterion S, Eq. (9.12), it is necessary that a non-vanishing adjoint vector function $[\mathbf{z}\,(t), z_t(t)]$ exists which satisfies adjoint Eq. (9.48) and transversality conditions (9.58)–(9.61), and that, for each time instant, the Hamiltonian H (or $\tilde{H}$) is maximum with respect to all controls, i.e., the optimal Hamiltonian $H(\widehat{\mathbf{z}}, \widehat{\mathbf{x}}, \widehat{\mathbf{u}}, t)$ satisfies the following equation

$$H(\widehat{\mathbf{z}}, \widehat{\mathbf{x}}, \widehat{\mathbf{u}}, t) = \max_{\mathbf{u} \in \mathrm{U}} H(\widehat{\mathbf{z}}, \widehat{\mathbf{x}}, \mathbf{u}, t) = M(\widehat{\mathbf{z}}, \widehat{\mathbf{x}}, t) \tag{9.72}$$

(symbols marked as $\widehat{\mathbf{z}}, \widehat{\mathbf{x}}, \widehat{\mathbf{u}}, etc$ pertain to an optimal process). Moreover, the maximum value of the enlarged Hamiltonian equals zero for every t, i.e.,

$$\tilde{H}\left(\widehat{\tilde{\mathbf{z}}}, \widehat{\tilde{\mathbf{x}}}, \widehat{\mathbf{u}}\right) = \max_{\mathbf{u} \in \mathrm{U}} \tilde{H}(\tilde{\mathbf{z}}, \tilde{\mathbf{x}}, \mathbf{u}) = \tilde{M}\left(\widehat{\tilde{\mathbf{z}}}, \widehat{\tilde{\mathbf{x}}}\right) = 0 \tag{9.73}$$

Note that the optimal control $\widehat{\mathbf{u}}$ is a function of enlarged state, i.e., $\widehat{\mathbf{u}}(t) = g(\widehat{\mathbf{z}}, \widehat{\mathbf{x}}, t)$. When the optimal control $\widehat{\mathbf{u}}$ lies in the interior of admissible region U and maximum of H or $\tilde{H}$ is of stationary nature, then the necessary optimality condition of criterion S with respect to controls $\mathbf{u}$ takes the form

$$\frac{\partial H}{\partial u_l} = \frac{\partial \tilde{H}}{\partial u_l} = 0. \tag{9.74}$$

In an autonomous case, when the process model does not contain time explicitly, the relationship $\dot{z}_t = -\partial H / \partial t = 0$, and then both z_t and $M = \max H$ are constant along the optimal path. If, in addition, time t_k is free, i.e., equations of Ω_k do not contain t_k explicitly, then $z_t = 0$ and $\max H = M = 0$.

It should also be kept in mind that adjoint variables may be defined differently in different works. In particular, new adjoints $\psi_I = -z_i$ and $\psi_t = -z_t$ can be accepted as alternative adjoint variables. Then, the Hamiltonian maximum appears for minimum of the performance criterion.

As *the first modification of basic maximum principle*, an algorithm of the process with parameters can be considered. State equations have then the form

$$\frac{d\mathbf{x}}{dt} = \mathbf{f}(\mathbf{x}, \mathbf{u}, \mathbf{p}, t) \quad t_p \leq t \leq t_k \tag{9.75}$$

where $\mathbf{p} = (p_1, p_2, \ldots, p_k)$ is the vector of parameters, i.e., quantities constant along the trajectory. Absence of constraints imposed on these parameters is assumed. In an optimization problem, it is required to find not only optimal control $\widehat{\mathbf{u}}(t)$ and optimal trajectory $\widehat{\mathbf{x}}(t)$, but also optimal (S-maximizing) level of each parameter ($p_1, p_2, \ldots, p_k$). Assumptions regarding boundary conditions, constraints, etc., (Section 9.2.1), remain unchanged.

In order to solve the optimization problem, equations of state (9.2) are supplemented by l subsidiary equations

$$\frac{d\mathbf{p}}{dt}=0 \quad \mathbf{p}=(p_1,p_2,\ldots,p_k) \tag{9.76}$$

and the state space $\mathbf{y}$ with coordinates $y_1=x_1$, $y_2=x_2$, $\ldots y_s=x_s$, $y_{s+1}=p_1$, $y_{s+2}=p_2, \ldots y_{s+k}=p_k$. Explicit time t can be treated as the coordinate y_{s+k+1}. For the new state space $\mathbf{y}$ the already-known optimization algorithm, Eqs. (9.47)–(9.50) is employed. In the space of variables y_l, the initial coordinates p_{jp} and the final coordinates p_{jk} are free because equations of manifolds Ω_p and Ω_k do not contain variables p_j. Since the quantities p_j do not appear in the function G, the transversality conditions for coordinates of vectors $\mathbf{p}(t_p)$ and $\mathbf{p}(t_k)$ are as follows:

$$z_{s+1}(t_p) = z_{s+2}(t_p) = \ldots = z_{s+k}(t_p) = 0 \tag{9.77}$$

$$z_{s+1}(t_k) = z_{s+2}(t_k) = \ldots = z_{s+k}(t_k) = 0 \tag{9.78}$$

Calculating the problem Hamiltonian in the state space of variables $\mathbf{y}$, we obtain

$$H_y = H(\mathbf{z}, y, \mathbf{u}, t) = \sum_{i=1}^{s} z_i f_i + f_0 + z_{s+1}{*}0 + z_{s+2}{*}0 = H \tag{9.79}$$

which shows that the Hamiltonian is the same as in the space without parameters. Similarly, for enlarged Hamiltonians, $\widetilde{H}_y=\widetilde{H}$. For the original adjoint variables $z_1, z_2 \ldots z_s$ Eqs. (9.48) or (9.53) are obtained in an unchanged form along with the supplementary equations

$$\frac{dz_{s+j}}{dt} = -\frac{\partial H}{\partial p_j} = -\left[\sum_{i=1}^{s} \frac{\partial f_i}{\partial p_j} + \frac{\partial f_0}{\partial p_j}\right] \mathrm{j} = 1,2,\ldots,k \tag{9.80}$$

Conditions (9.72) and (9.73), which describe maxima of Hamiltonians H and $\widetilde{H}$, modified by the presence of parameters, remain unchanged.

While all that is necessary to solve this problem is already obtained, an attempt to get the necessary optimality conditions in the form similar to the form without parameters is reasonable. Therefore, it is recommended to get rid of extra parameters $z_{s+1}, \ldots z_{s+k}$. Integrating for this purpose Eq. (9.80) between the limits t_p and t_k, and employing the transversality conditions (9.77) and (9.78), k equations is obtained,

$$\int_{t_p}^{t_k} \left[\sum_{i=1}^{i=s} z_i \frac{\partial f_i}{\partial p_j} + \frac{\partial f_0}{\partial p_j}\right] dt = 0 \quad \mathrm{j}=1,2,\ldots,k \tag{9.81}$$

which replace Eqs. (9.77), (9.78), and (9.80). Thus, supplementary adjoint variables z_{s+1}, ...z_{s+k} are eliminated. Since the integrand of Eq. (9.81) contains the partial derivative $\partial H/\partial p_j$, the following conclusion is valid:

For the optimality of a process with unconstrained parameters, all relationships derived for the maximum principle without parameters are valid, yet k supplementary equations must be satisfied.

$$\int_{t_p}^{t_k} \frac{\partial H}{\partial p_j} dt = 0 \quad j = 1,2,\ldots,k \tag{9.82}$$

which constitute optimality conditions of performance criterion S, Eq. (9.12), with respect to free parameters $\mathbf{p}=(p_1, p_2, \ldots, p_k)$. Boltyanski (1971) has worked out a generalization of this problem for constrained parameters $\mathbf{p}=(p_1, p_2, \ldots, p_k)$.

The *second modification of basic maximum principle*, generalizes the basic algorithm for processes with local constraints imposed on state coordinates, controls and time. Consequently, the optimization problem is one of the most difficult. An outline of the corresponding solving method is described below.

It is assumed that h inequality constraints

$$\phi_\alpha(\mathbf{x}, t) \leq 0 \quad j = 1,2,\ldots,k \tag{9.83}$$

is supplemented to the original optimization model formulated in Section 9.2.1. In the enlarged state space, equality $\phi_\alpha=0$ describes certain $(s+1\text{-}h)$ manifold. Thus, the condition (9.83) says that the optimal trajectory can be found only in a definite region of the state space $(\mathbf{x},t)$, or, when all constraints 9.83) are operative, the trajectory or its part reside on the manifold $\boldsymbol{\phi}=0$.

The corresponding change in the derivation of the maximum principle algorithm involves seeking the maximum of function $B(\mathbf{x}, \mathbf{u}, t)$ subject to constraints (9.83). Therefore, Eqs. (9.38), (9.39) or conditions $\partial B(\mathbf{x},t,\mathbf{u})/\partial x_i = 0$ and $\partial B(\mathbf{x},t,\mathbf{u})/\partial t = 0$, should be replaced by the generalized conditions:

$$\frac{\partial B(\mathbf{x}, t, \mathbf{u})}{\partial x_i} + \lambda_j \frac{\partial \phi_1}{\partial x_i} + \ldots. + \lambda_h \frac{\partial \phi_h}{\partial x_i} = 0 \tag{9.84}$$

$$\frac{\partial B(\mathbf{x}, t, \mathbf{u})}{\partial t} + \lambda_j \frac{\partial \phi_1}{\partial t} + \ldots. + \lambda_h \frac{\partial \phi_h}{\partial t} = 0 \tag{9.85}$$

resulting from applying the Kuhn-Tucker theorem of static optimization. The vector of Lagrange multipliers $\boldsymbol{\lambda}$ has the components λ_1, λ_2, ...λ_h. These components are determined as follows:

$$\lambda_\alpha = 0 \text{ for inoperative constraint } \phi_\alpha \tag{9.86}$$

$$\lambda_\beta \text{ is defined by the condition } \phi_\beta = 0 \text{ for operative constraint } \phi_\beta \tag{9.87}$$

In view of the relations (9.84) and (9.85), Eqs. (9.42), (9.43) are modified to.

forms which respectively contain extra terms $-\sum_{1}^{h} \lambda_i \partial\phi_i/\partial x_j$ and $-\sum_{1}^{h} \lambda_i \partial\phi_i/\partial t$ on the right hand sides of these equations. Consequently, the differential equations for adjoint variables z_i and z_t assume the form

$$\frac{dz_i}{dt} = -\frac{\partial H}{\partial x_i} - \sum_{j=1}^{h} \lambda_j \frac{\partial \phi_j}{\partial x_i} \quad i = 1,2,\ldots,s \tag{9.88}$$

$$\frac{dz_t}{dt} = -\frac{\partial H}{\partial t} - \sum_{j=1}^{h} \lambda_j \frac{\partial \phi_j}{\partial t} \tag{9.89}$$

Eqs. (9.88), (9.89) characterize basic modifications of maximum principle in case of constraints imposed on state variables. For more details, see literature (Pontryagin et al., 1962; Bryson et al., 1963; Bryson and Denham, 1964; Boltyanski, 1971; Findeisen et al., 1977).

9.3.2 Special cases and singular controls

Some special cases of optimization algorithm

When the performance criterion depends only on the final state:

$$S = \sum_{i=1}^{s} c_i x_i(t_k) \tag{9.90}$$

and the initial coordinates and final time are fixed, coordinates of the final state $x_i(t_k)$ may be free, and, as such, they may be chosen optimally. For the process optimality, canonical Eqs. (9.48), (9.50) and the maximum Hamiltonian Eq. (9.72) must be satisfied. Transversality condition has the form

$$z_i(t_k) = c_i \tag{9.91}$$

(compare the use of Eq. (9.59) for free end variables and note that the adjoint variable z_t need not be considered).

For the optimization problem in which some of the variables $x_i(t_k)$, (for example $x_a(t_k)$), are fixed, the optimization criterion can be written as

$$S = \sum_{i=1, i \neq a}^{s} c_i x_i(t_k) \tag{9.92}$$

and the basic algorithm is still valid, provided that Eq. (9.91) involves only coordinates $z_i(t_k)$ different than $z_a(t_k)$. The condition for the adjoint $z_a(t_k)$ is then not required; it is replaced by the

condition $x_a(t_k)=\bar{x}_{ak}$.
Let us assume that m of initial values x_{jp} are free and that S of Eq. (9.90) still has to be maximized. Then m conditions

$$z_t\left(t_p\right) = 0 \tag{9.93}$$

must appear in the maximum principle algorithm. When the final time t_k is free, then the condition $z_t(t_k)=0$ is equivalent to the vanishing of the final Hamiltonian

$$\widehat{H}(t_k) = 0 \tag{9.94}$$

This condition makes it possible to ignore the adjoint z_t in any part of optimization algorithm. Still, the use of adjoint z_t may be helpful in non-autonomous systems.

Singular controls

Eqs. (9.47) or (9.72), or other relations describing the maximum H condition, do not always define uniquely the optimal control $\widehat{\mathbf{u}}(t)$ within the entire range of times (t_p, t_k). When this non-uniqueness appears in a finite number of points, where control discontinuity appears, then the effective use of the maximum principle is still possible. However, a class of mathematical models exists where the maximum H condition does not uniquely define $\widehat{\mathbf{u}}(t)$ in a certain part of the time range (t_p, t_k) or even in the entire range (t_p, t_k). Controls of this sort are called *singular* (Athans and Falb, 1966). They arise, for instance, when in a certain time interval, H does not explicitly depend on control $\mathbf{u}$ because an adjoint variable vanishes in this interval. To exemplify the problem, let us consider the state equations

$$\frac{dx}{dt} = r(x, T) \tag{9.95}$$

$$\frac{dT}{dt} = Ar(x,T) + B(T_m - T) \quad A \equiv \frac{(-\Delta H)}{\rho c_p}; \quad B \equiv \frac{a_w F_w}{\rho c_p V}(T_m - T) \tag{9.96}$$

where x - product concentration, T- temperature in the reactor and T_{m} - temperature of external medium (cooling medium in case of an exothermal reaction). Eqs. (9.95), (9.96) describe a single chemical reaction in a batch reactor with external heat exchange. A and B are positive constants depending on specific heat c_p, reaction heat ΔH, and reactor geometry. The state variables are x and T, whereas T_m is the sole control. Process constraints are represented by the inequalities

$$T_{m*} \le T_m \text{ and } T_m \le T^{m*} \tag{9.96'}$$

The optimization criterion, S, which should be maximized,

$$S = x(t_k) \tag{9.97}$$

is the final concentration of the product. Note that S has the structure of Eq. (9.91) with $c_1=1$and $c_2=0$. The process Hamiltonian

$$H = z_1 r(x, T) + z_2 A r(x, T) + z_2 B(T_m - T) \tag{9.98}$$

is a linear function of control T_m. Therefore, it might be concluded that the optimal profile of coolant should be piecewise constant and assume only boundary values maximizing H, i.e.,

$$\widehat{T}_m = \begin{cases} T_{m*}\, if z_2 < 0 \\ \\ T_m^*\, if z_2 > 0 \end{cases} \tag{9.99}$$

This conclusion is insufficient, however, because it does not takes into account the possibility of singular controls, discussed below.

In general, in agreement with Eq. (9.99), optimal control $\widehat{T}_m(t)$ is the composition of the straight line segments T_{m*} and T_m^* as well as the curvilinear segment with the singular control, for which.

$$z_2 = 0 \tag{9.100}$$

It follows from Hamiltonian (9.98) and adjoint equation for $z_2=0$, that

$$\frac{dz_2}{dt} = -\frac{\partial H}{\partial T} = -\frac{\partial[z_1 r(x, T)]}{\partial T} = -z_1 \frac{\partial r}{\partial T} = 0 \tag{9.101}$$

which shows that whenever $z_1 \neq 0$ the derivative of reaction rate r versus temperature T must vanish. This result proves that, along the line segment with the singular control, the reaction rate must attain the stationary maximum with respect to T. The result, valid in the absence of operative constraints on T, agrees with the well-known conclusion in the theory of exothermal rectors. This conclusion might remain unnoticed if the possibility of singular controls was not considered. This example confirms, in particular, the appropriateness of Athans and Falb's (1966) call for systematic investigation of singular controls in optimally controlled systems.

9.4 Solving methods for maximum principle equations

Here is a brief summary of principles used when solving maximum principle equations, applied to the basic algorithm.

1. The optimization problem is broken down to the standard form making possible the use of the basic algorithm, Section 9.3.1, or some of its modifications. Defined are: state equations, controls, constraints, equations of initial and final manifolds, and the performance criterion.
2. Defined is the Hamiltonian function, and the problem of maximum H with respect to controls $\mathbf{u} \in \mathbf{U}$ is solved analytically. This yields the so-called extremal control in the form of function $\widehat{\mathbf{u}}(\mathbf{x}, \mathbf{z}, t)$. Whenever several local maxima of H exist, global maximum of H versus $\mathbf{u} \in \mathbf{U}$ needs to be selected in accordance with the maximum principle.

3. Optimal function $\widehat{\mathbf{u}}(\mathbf{x}, \mathbf{z}, t)$ is substituted into adjoint and state equations, which leads to a system of differential equations containing only state equations, x_i, adjoints z_i, and time, t.
4. The obtained system of differential equations is solved with the use of 2s+2 boundary conditions (initial and final). Usually, this is a complicated task, because of the so-called two-point boundary conditions (a part of boundary conditions pertains to the beginning and another part—to the end of the trajectory). Extremals $\mathbf{x}(t)$, $\mathbf{z}(t)$ and $z_t(t)$ are then found.
5. Equations of extremals are substituted into the function $\widehat{\mathbf{u}}(\mathbf{x}, \mathbf{z}, t)$, which yields the simplest function $\widehat{\mathbf{u}}(t)$ describing optimal control.

However, effective application of purely analytical methods is seldom possible; thus, many numerical methods have proposed to solve the maximum principle equations. Usually trial and error procedures are necessarily associated with satisfaction of boundary conditions and the use of the so-called extremal search approaches (Pontryagin et al., 1962; Athans and Falb, 1966; Bryson et al., 1963; Bryson and Denham, 1964; Boltyanski, 1971; Findeisen et al., 1977; Fan, 1966; Findeisen et al., 1977; Sieniutycz, 1991, and many others).

9.5 Discrete versions of maximum principle

9.5.1 Discrete algorithm with constant Hamiltonian

Refer now to a cascade composed of perfectly mixed stages, with time intervals θ^{n} and other controls $\mathbf{u}^{\mathrm{n}}$ as decisions at stage n. This cascade also schematizes the principle of multistage optimization calculations by the dynamic programming method (forward algorithm) in accordance with Bellman's principle of optimality (Bellman, 1967; Bellman and Dreyfus, 1962). On the other hand, Fig. 9.7 in this chapter, shows a generalized discrete scheme which illustrates the multistage operation with product recycle. Both schemes are examples of discrete operations to which a discrete optimization theory can be applied. Such a theory is outlined below.

Prior to conclusions obtained further, we stress that the maximum principle for processes described by difference equations (discrete systems) not always can be linked with the maximum of a corresponding Hamiltonian. Therefore, the terminology with the wording "maximum principle" is not exact in the discrete world, yet it is used because of the literature tradition.

Let us compare the multistage schemes of discrete systems shown in Fig. 9.7. They both represent cascades with ideal mixing at the stage. It is assumed that the process state at stage n is described by the state vector $\mathbf{x}^{\mathrm{n}} = (x_1^n, x_2^n, \ldots x_s^n)$, whereas the control at stage n - by the control vector $\mathbf{u}^{\mathrm{n}} = (u_1^n, u_2^n, \ldots u_r^n)$. Additional variables, of control type, which appear in discrete mathematical models, are time intervals $\theta^{\mathrm{n}} = t^{\mathrm{n}} - t^{\mathrm{n}-1}$. The time variable, t^n, used here, represents a continuous time in a broad sense. Thus, the variable t^n may represent, for example:

chronological time, residence time, or spatial time, depending on the physical situation. However, the mathematical model does not impose any local constraints on time intervals θ^n, although the integral constraints imposed on the sum of controls θ^n can appear. The analysis shows that for the optimality of the sequence $\boldsymbol{\theta} = (\theta^1, \theta^3, \ldots \theta^n \ldots \theta^N)$, with all θ^n free, an enlarged Hamiltonian $\tilde{H}^n$ must necessarily vanish. This is, of course, an analogue with the vanishing enlarged Hamiltonian in continuous systems.

Under usual assumption of ideal mixing at the stage, equations of the discrete process have the following form

$$x_1^n - x_1^{n-1} = \theta^n f_i^n\left(x_1^n, x_2^n, \ldots, x_s^n t^n, u_1^n, u_2^n, \ldots, u_r^n\right) \tag{9.102}$$

$$t_1^n - t_1^{n-1} = \theta^n \quad i = 1,2,\ldots,s \quad n = 1,2,\ldots,N \tag{9.103}$$

and the simplest boundary conditions, considered first, fix both end points of the trajectory (initial and final point), i.e.,

$$x_i^0 = \bar{x}_{ip}, \quad t^o = \bar{t}^p \tag{9.104}$$

$$x_i^n = \bar{x}_{ik}, \quad t^N = \bar{t}^k \tag{9.105}$$

The state Eqs. (9.102), (9.103) can be written in the vector form

$$\frac{\mathbf{x}^n - \mathbf{x}^{n-1}}{\theta^n} = \mathbf{f}^n\left(\mathbf{x}^n, t^n, \mathbf{u}^n\right) \tag{9.106}$$

$$\frac{t_1^n - t_1^{n-1}}{\theta^n} = 1 \tag{9.107}$$

$$\mathbf{x}^0 = \bar{\mathbf{x}}_p, \quad t^o = \bar{t}^p \tag{9.108}$$

$$\mathbf{x}^N = \bar{\mathbf{x}}_k, \quad t^N = \bar{t}^k \tag{9.109}$$

Here, $\mathbf{x}^n$ is n-dimensional vector describing states of a control process, which develops in continuous time t^n and in discrete time, n. Both n and t are monotonously increasing quantities. Symbol $\mathbf{u}^n$ denotes r-dimensional control vector representing subsequent decisions. This vector should usually satisfy certain constraints, i.e., lies in an admissible region $\boldsymbol{U}$ of the control space, meaning that $\mathbf{u}^n \in \boldsymbol{U}$, or

$$\psi_i^n\left(u_1^n, u_2^n, \ldots, u_r^n\right) \leq 0 \quad i = 1,2,\ldots,m \tag{9.110}$$

In rare cases, however, the admissible region can be defined by a system of m inequality constraints involving both controls and state coordinates

$$\psi_i^n\left(x_1^n, x_2^n, \ldots, x_s^n t^n, u_1^n, u_2^n, \ldots, u_r^n\right) \leq 0 \tag{9.111}$$

Since these constraints involve variables of two spaces (state space and control space) they make any solving procedure extremely complicated; therefore, we shall focus on the simpler constraints, Eq. (9.110).

The interval of the continuous time, $\theta^n = t^n - t^{n-1}$, is also a control variable, which, however, is considered separately. This distinction stresses the property of free θ^n sequences associated with the constancy of a Hamiltonian along the optimal discrete trajectory. However, this Hamiltonian is no longer constant when any active local constraints are imposed on θ^n.

The optimization problem is to find a decision sequence $\{\mathbf{u}^n = \mathbf{u}^1, \mathbf{u}^2, \ldots, \mathbf{u}^N\}$ which makes an optimization criterion an extremum (maximum or minimum) provided that all constraints imposed on state variables, controls, and time are satisfied.

As an optimization criterion a discrete Bolza functional S with gauging potential $G(\mathbf{x}, t)$ is accepted

$$S^N = \sum_{n=1}^{N} f_0(\mathbf{x}^n, t^n, \mathbf{u}^n)\theta^n + G(\mathbf{x}^N, t^N) - G(\mathbf{x}^0, t^0) \tag{9.112}$$

To derive optimality conditions, the forward algorithm of dynamic programming is applied (re: subscript f in the optimal performance function, Q_f^n). The state equations are expressed in the form of explicit functions of state coordinates $\mathbf{x}^n$ and time t^n

$$\mathbf{x}^{n-1} = \mathbf{x}^n - \theta^n \mathbf{f}^n(\mathbf{x}^n, \mathbf{u}^n, t^n) \quad t^{n-1} = t^n - \theta^n \tag{9.113}$$

Designating the optimal performance function

$$Q_f^n(\mathbf{x}^n, t^n) = \max_{\mathbf{u}^n, \theta^n} \left\{ \sum_{m=1}^{n} f_0^m(\mathbf{x}^m, t^m, \mathbf{u}^m)\theta^m + G^n(\mathbf{x}^n, t^n) - G^o(\mathbf{x}^0, t^0) \right\} \tag{9.114}$$

and applying Bellman's principle of optimality, the following recurrence equation (forward algorithm) is obtained

$$Q_f^n(\mathbf{x}^n, t^n) = \max_{\mathbf{u}^n, \theta^n} \left\{ \begin{array}{l} f_0^m(\mathbf{x}^m, t^m, \mathbf{u}^m)\theta^m \\ + G^n(\mathbf{x}^n, t^n) - G^{n-1}(\mathbf{x}^{n-1}, t^{n-1}) + Q_f^n(\mathbf{x}^{n-1}, t^{n-1}) \end{array} \right\} \tag{9.115}$$

which applies Eq. (9.113). Further transformations use the optimal function

$$P^n(\mathbf{x}^n, t^n) \equiv Q_f^n(\mathbf{x}^n, t^n) - G^n(\mathbf{x}^n, t^n) \tag{9.116}$$

Note a similar function, $R(\mathbf{x}, t)$, Eq. (9.29), used in the backward DP algorithm of the continuous problem. In this section, however, the forward DP algorithm, which deals with profit function evaluated in terms of outlet states, $Q^n(\mathbf{x}^n, t^n)$, is applied. Using the definition of function P^n and the state equations to eliminate variables $(\mathbf{x}^{n-1}, t^{n-1})$from Eq. (9.115) and transferring the function $Q_f^n(\mathbf{x}^n, t^n)$ to the inside of maximizing bracket in Eq. (9.115) yields

$$\max_{\mathbf{u}^n \in U, \theta^n} \{f_0^n(\mathbf{x}^n, t^n, \mathbf{u}^n)\theta^n + P^{n-1}(\mathbf{x}^n - \theta^n \mathbf{f}^n, t^n - \theta^n) - P^n(\mathbf{x}^n, t^n)\} = 0 \tag{9.117}$$

In order to employ the observation that Eq. (9.115) is valid for arbitrary values of variables $\mathbf{x}^n$ and t^n the following function is considered

$$A^n(\mathbf{x}^n, t^n, \mathbf{u}^n, \theta^n) \equiv f_0^n \theta^m + P^{n-1}(\mathbf{x}^n - \theta^n \mathbf{f}^n, t^n - \theta^n) - P^n(\mathbf{x}^n, t^n) \tag{9.118}$$

It is assumed that A^n is differentiable with respect to all state coordinates and time. The function $A^n(\mathbf{x}^n, t^n, \mathbf{u}^n, \theta^n)$ satisfies the following conditions

$$A^n(\mathbf{x}^n, t^n, \mathbf{u}^n, \theta^n) \leq 0 \quad \text{for all } \mathbf{x}^n, t^n, \mathbf{u}^n, \text{ and } \theta^n \tag{9.119}$$

$$A^n(\mathbf{x}^n, t^n, \mathbf{u}^n, \theta^n) = 0 \text{ for every optimal process } \hat{\mathbf{x}}^n, \hat{t}^n, \hat{\mathbf{u}}^n, \text{ and } \hat{\theta}^n \tag{9.120}$$

This equality means that, for an arbitrary point ($\mathbf{x}^n$, t^n), a control $\mathbf{u}^n \in U$ and an interval θ^n can be found such that the equality $A^n(\mathbf{x}^n, t^n, \mathbf{u}^n, \theta^n) = 0$ holds.

Let us fix certain $\mathbf{u}^n$ in the admissible region U. Then, it follows from Eqs. (9.119), (9.120) that function $A^n(\mathbf{x}^n, t^n, \mathbf{u}^n, \theta^n)$ achieves the maximum also with respect to variables x_i^n and t^n, thus the partial derivatives $\partial A^n / \partial x_i^n$ and $\partial A^n / \partial t^n$, must vanish, i.e.,

$$\frac{\partial f_0^n}{\partial x_i^n}\theta^n - \sum_{j=1}^{s} \frac{\partial P^{n-1}}{\partial x_j^{n-1}} \frac{\partial f_j^n}{\partial x_i^n}\theta^n + \frac{\partial P^{n-1}}{\partial x_i^{n-1}} - \frac{\partial P^n}{\partial x_i^n} = 0 \quad i = 1, 2, \ldots, s \tag{9.121}$$

$$\frac{\partial f_0^n}{\partial t^n}\theta^n - \sum_{j=1}^{s} \frac{\partial P^{n-1}}{\partial x_j^{n-1}} \frac{\partial f_j^n}{\partial t^n}\theta^n + \frac{\partial P^{n-1}}{\partial t^{n-1}} - \frac{\partial P^n}{\partial t^n} = 0 \quad i = 1, 2, \ldots, s \tag{9.122}$$

This result holds whenever constraints on the process trajectory are absent. Assuming the stationary nature of optimal controls $\mathbf{u}^n$ and θ^n and calculating partial derivatives of A^n with respect to controls $\mathbf{u}^n$ and θ^n as well as adding Eq. (9.117) to the obtained equations with omitted minimization sign, the following set of equations is obtained

$$\frac{\partial f_0^n}{\partial u_l^n} - \sum_{j=1}^{s} \frac{\partial P^{n-1}}{\partial x_i^{n-1}} \frac{\partial f_i^n}{\partial u_l^n} = 0 \tag{9.123}$$

$$f_0^n - \sum_{j=1}^{s} \frac{\partial P^{n-1}}{\partial x_i^{n-1}} f_i^n - \frac{\partial P^{n-1}}{\partial t^{n-1}} = 0 \tag{9.124}$$

$$P^n(\mathbf{x}^n, t^n) - P^{n-1}\left(\mathbf{x}^{n-1}, t^{n-1}\right) = f_0^n \theta^n \tag{9.125}$$

This set describes, together with Eqs. (9.121), (9.122), the necessary optimality condition of the discrete process considered.

After defining adjoint variables as following partial derivatives

$$z_i^{n-1} = -\frac{\partial P^{n-1}(\mathbf{x}^{n-1}, t^{n-1})}{\partial x_i^{n-1}} \tag{9.126}$$

$$z_t^{n-1} = -\frac{\partial P^{n-1}(\mathbf{x}^{n-1}, t^{n-1})}{\partial t^{n-1}} \tag{9.127}$$

and the Hamiltonian function as

$$H^{n-1}\left(\mathbf{x}^n, t^n, \mathbf{z}^{n-1}, \mathbf{u}^n\right) = f_0^n(\mathbf{x}^n, t^n, \mathbf{u}^n) + \sum_{j=1}^{s} z_i^{n-1} f_i^n(\mathbf{x}^n, t^n, \mathbf{u}^n) \tag{9.128}$$

Eq. (9.124) takes the form

$$z_t^{n-1} + H^{n-1}\left(\mathbf{x}^n, t^n, \mathbf{z}^{n-1}, \mathbf{u}^n\right) = 0. \tag{9.129}$$

This is a discrete Hamilton-Jacobi-Bellman equation, a counterpart of the continuous HJB Eq. (9.47). In terms of the Hamiltonian, equations of state (9.106) and (9.107) and necessary optimality conditions (9.121), (9.122), and (9.123) can be written in terms of H^{n-1} as follows

$$\frac{x_i^n - x_i^{n-1}}{\theta^n} = \frac{\partial H^{n-1}}{\partial z_i^{n-1}} \tag{9.130}$$

$$\frac{t^n - t^{n-1}}{\theta^n} = \frac{\partial H^{n-1}}{\partial z_t^{n-1}} \tag{9.131}$$

$$\frac{z_i^n - z_i^{n-1}}{\theta^n} = -\frac{\partial H^{n-1}}{\partial x_i^n} \tag{9.132}$$

$$\frac{z_t^n - z_t^{n-1}}{\theta^n} = -\frac{\partial H^{n-1}}{\partial t^n} \tag{9.133}$$

and, for a stationary optimum

$$\frac{\partial H^{n-1}}{\partial u_j^n} = 0 \tag{9.134}$$

It follows that the definitions of adjoint variables in continuous and discrete version are consistent in preserving the same sign of both adjoints. Eqs. (9.129)–(9.134) are analogous to corresponding equations of continuous processes. It should be underlined, however, that condition (9.134) has in the discrete theory the different meaning than in the continuous theory. In continuous processes, the vanishing of derivatives $\partial H^{n-1}/\partial u_j^n$ is satisfied for the stationary maximum of H^{n-1}, whereas in discrete ones the condition (9.134) is only necessary for a stationary maximum of function A^n, Eq. (9.118), but it cannot be said that this maximum must imply the maximum of discrete H^{n-1}, Eq. (9.118), versus controls $\mathbf{u}^n$. Indeed, considering only Eq. (9.134) it is insufficient to conclude the maximum H^{n-1}, because vanishing of $\partial H^{n-1}/\partial u_j^n$

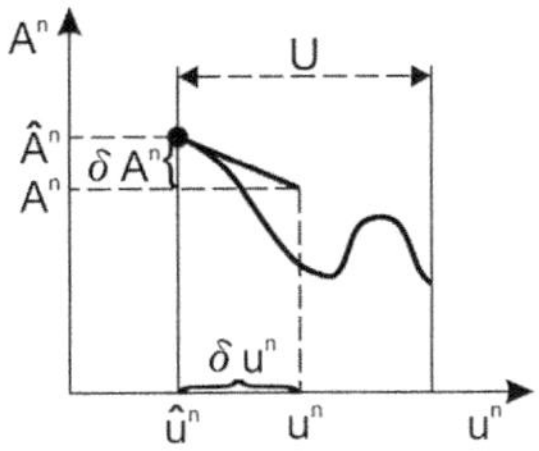

Fig. 9.6
Interpreting maximum condition for function A^n in one-dimensional space $\mathbf{u}^n$, (Sieniutycz, 1991, his condition (9.136)).

can imply equally well the local maximum, the local minimum or the saddle of the Hamiltonian H^{n-1}. Therefore, it is more appropriate to call the discrete algorithm, Eqs. (9.129)–(9.134), the principle of the stationary discrete Hamiltonian, rather than the discrete maximum principle.

Additional conclusions can be drawn from a consideration of the local boundary maximum of function A^n, Eq. (9.118). In this case, the necessary maximum condition is

$$\frac{\partial A^n}{\partial \widetilde{\mathbf{u}}^n}\delta \mathbf{u}^n \leq 0 \quad \delta \mathbf{u}^n = \mathbf{u}^n - \widetilde{\mathbf{u}}^n \tag{9.135}$$

for sufficiently small $\delta \mathbf{u}^n \in U$. Interpretation of this inequality is given in Fig. 9.6.

Applying Eq. (9.118) in Eq. (9.134) or Eq. (9.34), calculating the derivative $\partial A^n/\partial \tilde{\mathbf{u}}\, n$ and employing the definition of the adjoint variables and Hamiltonian, yields

$$\theta^n \frac{\partial H^{n-1}}{\partial \widetilde{\mathbf{u}}^n}\delta \mathbf{u}^n \leq 0 \tag{9.136}$$

Since the time intervals θ^n are positive, inequality (9.136) proves that the discrete Hamiltonian H^{n-1} has a local maximum versus controls $\mathbf{u}^n$ whenever the optimal control lies at the boundary of the admissible regime U. This result is weaker than the corresponding result in the continuous version, where the optimal control ensures the global maximum of the Hamiltonian. If, however, the interval θ^n decreases, then the magnitude of variation $\delta \mathbf{u}^n$ satisfying inequality (9.136) increases, so that in the limit when $\theta^n \to 0$ the variation comprises the whole regime U. Inequality (9.136) then becomes satisfied for every $\mathbf{u}^n \in \mathbf{U}$, which proves that the local maximum becomes the global one, and this is the content of the maximum principle for continuous processes.

The above conclusions make it possible to formulate the discrete maximum principle in the form of the following theorem:

Optimal control $\widehat{\mathbf{u}}^n$ belonging to the boundary of the admissible regime U ensures a local maximum of the Hamiltonian H^{n-1}, whereas the optimal control belonging to the interior of the admissible regime U ensures a stationary point of the Hamiltonian.

Several words should be devoted to the transversality conditions, which arise when simple boundary conditions (9.108) and (9.109) do not hold. Rather, it is required that the initial and final points of the discrete trajectory reside on certain manifolds, Ω_p and Ω_k. Considering the variation of functional S caused by the free motion of the right end of an extremal, we find

$$\delta S_k^N = \delta Q_f^N = \sum_{i=1}^{s} \frac{\partial Q_f^N}{\partial x_i^N} \delta x_i^N + \frac{\partial Q_f^N}{\partial t} \delta t^N \tag{9.137}$$

When the definition of adjoint variables, Eqs. (9.126), (9.137), is employed, and the equality $Q_f^N = P^N + G^N$ is used, the variation of S resulting from the motion of both ends of the extremal is

$$dS^N = \sum_{i=1}^{s} \left(\frac{\partial G}{\partial x_i} - z_i \right) \delta x_i + \left(\frac{\partial G}{\partial t} - z_t \right) \delta t \Bigg|_{\substack{x_i^0, t^0 \\ z_i^0, z_t^0}}^{\substack{x_i^N, t^N \\ z_i^N, z_t^N}} \tag{9.138}$$

This result is analogous to that obtained for the continuous version despite of some intermediate definitions and equations. Eq. (9.137) along with Eqs. (9.60), (9.61) make it possible, for admissible variations on Ω_p and Ω_k, to write down the $Q_f^n(\mathbf{x}^n, t^n)$ transversality conditions for boundary conditions (9.13) and (9.14). In particular, transversality conditions for free coordinates and time follow. For example, final transversality conditions (for the right end of the trajectory) are of the form

$$z_j^N = \frac{\partial G^N}{\partial x_j^N} \tag{9.139}$$

$$z_t^N = -\widehat{H}^N = \frac{\partial G^N}{\partial t^N} \tag{9.140}$$

For a process autonomous versus time t^n, optimality conditions (9.129) and (9.133) yield the equality

$$\widehat{H}^n = \widehat{H}^{n-1} = constant \tag{9.141}$$

which describe the constancy of the discrete Hamiltonian along the optimal trajectory.

It should be kept in mind that the constancy of the discrete H in autonomous systems holds only for stationary optimal intervals θ^n. For arbitrary intervals θ^n, this constancy does not hold, yet in every optimal process constancy of time adjoint z_t^n is ensured.

Algorithms using Pontryagin's type Hamiltonians can also be applied in optimization of complex systems of non-sequential topology, (Fan, 1966; Szwast, 1988; Sieniutycz, 1991, and others).

Example: A discrete problem of minimum time.

Consider the first-order reaction $A \underset{k_2,E_2}{\overset{k_1,E_1}{\rightleftarrows}} B$ in an isothermal cascade of reactors with ideal mixing at the stage. Derive the sole state equation

$$\frac{x^n - x^{n-1}}{\theta^n} = x_e - x^n, \tag{9.142}$$

where x is the mass fraction of product B, and x_e is the equilibrium mass fraction. Solve the problem of the minimum total residence time $\Sigma\theta^n$ for this cascade. Apply the discrete optimization algorithm with the constant Hamiltonian H^n. Show the equality of optimal stage times θ^n.

9.5.2 Common algorithm of discrete maximum principle and modifications

However, the earliest versions of the discrete maximum principle do not attempt to establish Hamiltonians of Pontryagin's type nor to set optimization algorithms preserving as many as possible substantial properties of classical PMP in discrete cases. The related discussions are available in the literature (Fan and Wang, 1964); Halkin, 1966; Boltyanski, 1973; Findeisen et al., 1977; Sieniutycz, 1991, and others). Below the popular algorithm of Fan and Wang (1964) is outlined, as applied in the context of multistage operations with recirculation, Fig. 9.7.

State transformations at stage n are governed by a vector transformation function $\mathbf{T}^n$. The set of state equations is represented by a vector difference equation

$$\mathbf{x}^n = \mathbf{T}^n\left(\mathbf{x}^{n-1}, \mathbf{u}^n\right) n = 1,2,\ldots s \tag{9.143}$$

The mixing of a fresh stream (subscript p) and recirculating stream (subscript r) is described by a vector equation

$$\mathbf{x}^0 = \mathbf{M}^n\left(\mathbf{x}_p, \mathbf{x}^{\mathrm{N}}, q_p, q_r\right) \tag{9.144}$$

Quantities p and q are generally constant; from the mass conservation law, q^0 is the sum of q^p and q^r. The optimization problem is to find a sequence of controls u^n, $n=1$, 2..N, which maximize the performance criterion

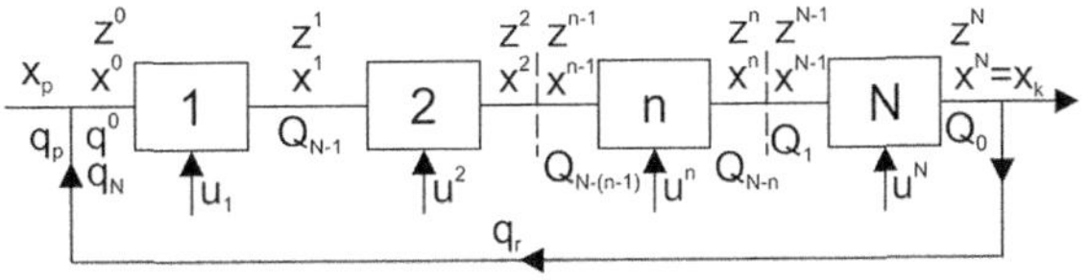

Fig. 9.7
Multistage operation with product recycling.

$$S = \sum_{i=1}^{s} c_i x_i^N \tag{9.145}$$

in which all final coordinates x_i^N are free. It is assumed that all initial coordinates of the fresh stream are fixed.

A solving procedure by the "discrete maximum principle" relies on introducing s-dimensional adjoint vector $\mathbf{z}^n$ and a Hamiltonian H^n satisfying the vector equations

$$H^n = \sum_{i=1}^{s} z_i^n T_i^n \left(\mathbf{x}^{n-1}, \mathbf{u}^n\right) n = 1,2\ldots,N \; i = 1,2,\ldots s \tag{9.146}$$

The Hamiltonian H^n can be remembered as the product $\mathbf{z}^n\mathbf{x}^n$. Definitions of H^n and state adjoints are not Pontryagin's. The vector equation for adjoint variables is of the form

$$\mathbf{z}^{n-1} = \frac{\partial H^n}{\partial \mathbf{x}^{n-1}} i = 1,2\ldots,s \tag{9.147}$$

A sequence of extremal controls for an interior maximum follows from the conditions

$$\frac{\partial H^n}{\partial \mathbf{u}^n} = 0 j = 1,2\ldots,r \tag{9.148}$$

Clearly, the Hamiltonian and adjoint variables of the present model differ from those of Pontryagin's algorithm. The present Hamiltonian does not have energy properties, and it is generally not constant along the discrete trajectory. Nonetheless, Sieniutycz (1991) offers a proof of the validity of the algorithm by the dynamic programming method. He also describes modifications and extensions of the algorithm.

9.5.3 Computational procedure

All numerical procedures require using iteration techniques and are applied regardless whether the Hamiltonian maximum is stationary or lies at the boundary of admissible region, U.

In a certain procedure, unknown coordinates at the end of the trajectory are assumed and verified. Subsequent steps in this procedure are as follows:

1. Assume coordinates of all vectors $\mathbf{x}^N$ and $\mathbf{z}^N$ which are not prescribed
2. Maximize Hamiltonian (9.145) at stage N with respect to controls. Determine nature of all stationary points satisfying the condition $\partial H^N/\partial u_j^N = 0$.
3. Employ control values u_j^N, obtained above, to calculate $\mathbf{z}^{N-1}$ and $\mathbf{x}^{N-1}$.
4. Repeat steps 2 and 3 for each stage until the first stage is achieved.
5. If the numerical values calculated for coordinates of vectors $\mathbf{x}^0$ and $\mathbf{z}^0$ satisfy Eq. (9.143) and other equations for final constraints, then the optimization problem is solved. Otherwise, the procedure should be repeated until an error function test will show that required accuracy is achieved.

Fan and Wang (1964), Boltyanski (1973) and Sieniutycz (1991) advocate other suitable procedures to solve numerically algorithms of discrete maximum principle.

9.6 Classification and comparison of computational methods for optimization

In the optimization of function of several variables, basic and special methods are distinguished among the optimization methods of functionals. Basic methods are both the most important and the most popular; for instance, all computational methods of solving maximum principle equations belong to the group of basic methods.

Basic methods can be divided into direct and indirect ones. In direct methods, monotonously optimizing sequences of admissible solutions are created, and the basis for acceptance of the next approximate solution is an improvement of its quality. Appropriate examples are: the Ritz method in the variational calculus (Elsgolc, 1960), dynamic programming (Bellman, 1967; Aris, 1964; Findeisen et al., 1977), the gradient methods in control spaces, method of second variation, and the method of variable operator in the control space. The two latter methods (Findeisen et al., 1977) are the most effective computational tools for the optimization of functionals. This is not surprising, because the method of second variation is a counterpart of Newton's classical method of function optimization (Sieniutycz, 1991) adapted in the functional control space. Similarly, a counterpart of the variable operator method in the function optimization theory is the method of Davidon (1959). In fact, the methods of Newton and Davidon are both fast convergent methods in function optimization, and this conclusion can be extended to optimization of functionals. Basic methods are also applied in problems in which except state equations additional constraints appear imposed on control and state variables (e.g., integral constraints or local constraints). A universal method of treating these constraints introduces the so-called penalty functionals and extremizes suitably modified optimization criteria (Findeisen et al., 1977).

Indirect basic methods determine the optimization solution with the help of necessary optimality conditions, for example, Euler-Lagrange equations and maximum principle equations. This group involves various methods of solving differential equations for extremals with (usually) two-point boundary conditions. One of these solving methods is Newton's method in the state space which deals with homogeneous equations for variations δx and δz obtained via linearizing the maximum principle equations. Yet, this method can occasionally be divergent as it may introduce difficulties following from instabilities of the adjoint equations in the natural direction of time. Usually, more effective are other methods based on the multiple solving of extremal equations which determine minimum of the error function dependent on the assumed initial conditions for adjoint variables (or the assumed final conditions for state variables). The convergence rate of such methods depends on the convergence rate of accepted optimization method of function of s independent variables. Such an optimization problem, which involves optimization of a function (instead of a functional), is usually characterized by much lower dimensionality than a corresponding problem appearing in other methods of functional optimization. Therefore, the convergence rate in the search methods is as a rule much faster than the convergence rate in other methods. In particular, the search methods are recommended for solving linear canonical equations. However, they may be ineffective when solving state and adjoint equations in the same direction of time, which often occurs for discrete problems.

There are also several special methods for functional optimization. Approaches applying penalty functionals to absorb complex state constraints belong to most effective special methods. By using penalty functionals, these methods abandon the standard way of treatment of constraints. Nonetheless, they are effective for very complicated (e.g., partial) differential equations of state. Into the family of special methods are often included methods employing specific (e.g., linear) forms of constraints for definite (e.g., discrete) state equations, and also multilevel methods.

References

Aris, R., 1964. Discrete Dynamic Programming. Blaisdell, New York.

Athans, M., Falb, P.L., 1966. Optimal Control–An Introduction to the Theory and its Applications. McGraw-Hill Book, New York.

Bellman, R., Dreyfus, S., 1962. Applied Dynamic Programming. Princeton University Press, Princeton.

Bellman, R.E., 1967. Introduction to Mathematical Theory of Control Processes. Academic Press, New York.

Boltyanski, V.G., 1971. Mathematical Methods of Optimal Control. Holt, Reihart and Wilson, New York.

Boltyanski, V.G., 1973. Optimal Control of Discrete Systems. Nauka, Moscow.

Bryson, A.E., Denham, W.F., 1964. Optimal programming problems with inequality constraints II. Solution by steepest-ascent. AIAA J. 2 (1), 25–34.

Bryson, A.E., Denham, W.F., Dreyfus, S.E., 1963. Optimal programming problems with inequality constraints I. Necessary conditions for extremal solutions. AIAA J. 1 (11), 2544–2550.

Davidon, W.C., 1959. Variable Metric Method of Minimization. A.E.C. Research and Development Report ANL-5990.

Elsgolc, L.E., 1960. Variational Calculus. Pańistwowe Wydawnictwa Naukowe, Warsaw.

Fan, L.T., 1966. The Continuous Maximum Principle: A Study of Complex System Optimization. Wiley, New York.

Fan, L.T., Wang, C.S., 1964. The Discrete Maximum Principle: A Study of Multistage System Optimization. Wiley, New York.

Findeisen, W., Szymanowski, J., Wierzbicki, A., 1977. Theory and Computational Methods of Optimization. Pańistwowe Wydawnictwa Naukowe, Warszawa (in Polish).

Gadewar, S.B., Doherty, M.F., Malone, M.F., 2001. A systematic method for reaction invariants and mole balances for complex chemistries. Comput. Chem. Eng. 25, 1199–1217.

Halkin, H., 1966. A maximum principle of the Pontryagin type for systems described by nonlinear difference equations. SIAM J. Control 4, 528–547.

Leitman, G., 1966. An Introduction to Optimal Control. McGraw-Hill, New York.

Leitman, G., 1981. The Calculus of Variations and Optimal Control. Plenum Press, New York.

Pontryagin, L.S., Boltyanski, V.G., Gamkrelidze, R.V., Mishchenko, E.F., 1962. The Mathematical Theory of Optimal Processes. Interscience Publishers, New York.

Sieniutycz, S., 1991. Optimization in Process Engineering, second ed. Wydawnictwa Naukowo Techniczne, Warszawa (in Polish).

Szwast, Z., 1988. Enhanced version of a discrete algorithm for optimization with a constant Hamiltonin. Inz. Chem. Proc. 3, 529–545.

Glossary

A	system ascendancy, Eq. (3.3)
$\mathbf{A}_{\mathbf{class}}$	classical affinity of reaction (J, $\mathrm{kJ\,kg^{-1}}$)
A	vector of driving affinities ($\mathrm{J\,mol^{-1}}$)
Anthr	anthropological age (years), Fig. 4.1
A_i	amount of the *i*th component in a reacting mixture (mol, kmol, kg)
A, B, C, D	parameters characterizing the behavior of food webs in Chapters 5–7
a_1, a_2, b_1, b_2, p, q	coefficients of the Lotka-Volterra model (–), Fig.1.1
a	nondimensional activity of the catalytic system (–)
a	power exponent (–)
B	exergy content in the system (J, kJ)
C	vector of molar concentrations in the system ($\mathrm{mol\,m^{-3}}$)
C	control parameter along the direction of the horizontal axis, Fig.1.2
c_i	molar concentration of the *i*th component ($\mathrm{mol\,m^{-3}}$)
D	effective diffusion coefficient ($\mathrm{m^2\,s^{-1}}$)
E	total energy or energy type quantity (J)
E	electric field vector ($\mathrm{kg\,m\,A\,s^{-3}}$)
F	area perpendicular to the flow ($\mathrm{m^2}$)
$F^n(x^n)$	optimal performance function of dynamic programming
$f_1, f_2 \ldots$	rate functions in the Pontryagin maximum principle
G	gauging function in Eq. (9.12) (–)
G	gas mass flux ($\mathrm{kg\,s^{-1}}$), molar flux ($\mathrm{mol\,s^{-1}}$)
g	gravity acceleration ($\mathrm{m\,s^{-2}}$)
H	various optimization Hamiltonians in Chapter 9
$H(\mathbf{x}, \mathbf{u}, \mathbf{p}, t)$	Hamiltonian function of the continuous process
$H^{n-1}(\mathbf{x}^n, \mathbf{u}^n, \mathbf{p}^{n-1}, t^n)$	Hamiltonian function of the discrete process
$\mathcal{H}$	nondimensional coefficient of homeostasis (–)

i	molar flux density of electric current ($mol\,m^{-1}\,s^{-1}$)
j	number of inequality constraints in Eq. (9.4)
K	chemical equilibrium constant (–)
k	reaction rate constant for chemical reaction ($mol\,s^{-1}$ for the first-order reaction)
k_{eff}	ka effective rate constant ($mol\,s^{-1}$ for the first-order reaction)
k	thermal conductivity coefficient ($W\,K^{-1}\,m^{-1}$)
k_B	Boltzmann constant ($1.381 \times 10^{-23}\,J\,K^{-1} = 1.3803 \times 10^{-16}$ $erg\,K^{-1}$)
l	length coordinate (m)
M_i	molar mass of the ith species ($kg\,kmol^{-1}$)
m	mass of the body, mass of particles (g, kg)
N	number of moles, number of particles (–)
N	total number of stages, number of transfer units (–)
N_A	Avogadro number $= 6.024 \times 10^{23}$ (mol^{-1})
n	number of states (–), number density (m^{-3})
p	horizontal parameter of the Lotka-Volterra model, Fig.1.1
p	probability distribution of a population in the Pielou (1969) model
q	vertical parameter of the Lotka-Volterra model, Fig.1.1
R	universal gas constant $= 8.3144598\,J\,K^{-1}\,mol^{-1}$ ($1.985\,cal\,K^{-1}$)
Ra	nondimensional Rayleigh number (–)
RAF	autocatalytic set, Fig. 2.2
r_j	rate of the jth reaction ($mol\,m^{-3}\,s^{-1}$)
S	entropy ($J\,K^{-1}$), information-theoretic entropy (–)
S_σ	entropy source or total dissipated entropy ($J\,K^{-1}$)
s	specific entropy or entropy per mass unit ($J\,K^{-1}\,g^{-1}$)
t	time or space-time variables (sec, h)
(u,v,w)	components of the affine vector
$V^n(x^n, t^n)$	discrete optimal profit function
(x,y,z)	components of vector in metric space
*	modified quantity mark
$v\,\rho^{-1}$	specific volume ($m^3\,kg^{-1}$)
X_1, X_2	state variables of the Lotka-Volterra model (–), Fig. 1.1
x $(x_1, x_2 \ldots x_i \ldots x_s)$	state vector of the general dynamical process
x $(x_1, x_2 \ldots x_i \ldots x_s, t)$	enlarged state vector of the general dynamical process
z	state adjoint variable

Greek symbols

a	overall heat transfer coefficient ($J\,m^{-2}\,s^{-1}\,K^{-1}$)
ϕ	electric potential (V)
Γ	complexity, complexity potential (–)
Δ	disequilibrium correction, increment, deviation (–)
η	first-law thermal efficiency $p(q_1)^{-1}$ (–)
λ	heat conductivity ($J\,m^{-1}\,K^{-1}\,s^{-1}$)
Θ	destruction rate of classical exergy ($J\,s^{-1}$)
θ	interval of the time type variable (s, –)
θ^n	time interval at stage n of a multistage process (s, –)
μ^k	chemical potential of the kth component ($J\,g^{-1}$)
ν	stoichiometric coefficient (–)
ω	frequency constant (s^{-1})
ρ	mass density ($kg\,m^{-3}$),
ρ_e	energy density ($J\,m^{-3}$)
ρ_s	entropy density ($J\,K^{-1}\,m^{-3}$)
σ	entropy production ($J\,K^{-1}\,s^{-1}$)
σ_s	density of entropy production ($J\,K^{-1}\,m^{-3}\,s^{-1}$)
$\tau = \omega t$	nondimensional time based on frequency constant (–)
τ	nondimensional time as the number of transfer units $xHTU^{-1}$ (–)
τ_{ik}	viscous stress (Pa)
Ω	order variable (–)
∇	differential operator

Subscripts

b	backward
c	cold
e	energy
f	fluid
g	gas bulk
h	hot
i	ith state variable
in	incoming
j	reaction number
k	kth component
m	molar quantity
mix	mixing
out	outcoming

p	product
s	solid
w	water
v	per unit volume
0	inlet state, reference state
1, 2	first and second, initial and final

Superscripts

a	power exponent
e	environment
f	final state and time
i	initial state and time
int	internal effect
n	stage number
T	transpose matrix, transform
′	modified variable, modified quantity
0	ideal state, equilibrium state

Abbreviations and acronyms

AIC	algorithmic information content
CC	computational complexity
CCS	complex computer simulations
CIT	classical irreversible thermodynamics
DP	dynamic programming
EIT	extended irreversible thermodynamics
FTT	finite-time thermodynamics
GA	genetic heredity of A
GEM	Gibbs equilibrium manifold
HJ	Hamilton-Jacobi equation
HJB	Hamilton-Jacobi-Bellman equation
LP	linear problem and linear programming
MEN	mass exchange network
Ni	mole number for the *i*th component in the mixture
NEE	rate of mole exchange between ecosystems
NLP	nonlinear problem/programming
OCT	optimal control theory

Index

Note: Page numbers followed by *f* indicate figures and *t* indicate tables.

M

N

O

P

9780443192371